PROBABILITY AND STATISTICS
THE SCIENCE OF UNCERTAINTY

THE HISTORY OF MATHEMATICS

PROBABILITY AND STATISTICS

THE SCIENCE OF UNCERTAINTY

John Tabak, Ph.D.

Checkmark Books®
An imprint of Facts On File, Inc.

PROBABILITY AND STATISTICS: The Science of Uncertainty

Checkmark Books
An imprint of Facts On File, Inc.
132 West 31st Street
New York NY 10001

Library of Congress Cataloging-in-Publication Data
Tabak, John.
    Probability and statistics : the science of uncertainty / John Tabak.
        p. cm. — (History of mathematics)
    Includes bibliographical references and index.
    ISBN 0-8160-4956-4 (hc) — 0-8160-6231-5 (pbk)
    1. Probabilities—History. 2. Mathematical statistics—History. I. Title.
    QA273.T23 2004
    519.2'09—dc222003016966

Checkmark Books are available at special discounts when purchased in bulk
quantities for businesses, associations, institutions, or sales promotions. Please call
our Special Sales Department in New York at (212) 967-8800 or (800) 322-8755.

You can find Facts On File on the World Wide Web at http://www.factsonfile.com

Text design by David Strelecky
Cover design by Kelly Parr
Illustrations by Patricia Meschino

Printed in the United States of America

MP FOF 10 9 8 7 6 5 4 3 2 1

This book is printed on acid-free paper.

To Gail with love.

CONTENTS

ACKNOWLEDGMENTS

The author is deeply appreciative of Frank Darmstadt, executive editor, for his many helpful suggestions and of Dorothy Cummings, project editor, for the insightful work that she did editing this volume.

Special thanks to Penelope Pillsbury and the staff of the Brownell Library, Essex Junction, Vermont, for their extraordinary help with the many difficult research questions that arose during the preparation of this book.

INTRODUCTION

WHAT IS PROBABILITY?

Some types of mathematics are as old as the written word. Geometry is probably older. We have written records from cultures that arose early in world history in which we can find geometry problems and solutions. Ancient scholars were working on geometry even as they were developing a system of writing to express their geometric insights. We can also point to cultures that never developed their own written language but did develop their own arithmetic. At least in some cultures, 'rithmetic is the oldest of the three Rs.

Arithmetic and geometry are branches of knowledge that concern aspects of our perception—number and shape—that are both abstract and concrete. For example, three birds and three books are both concrete manifestations of the abstract idea of the number 3. Similarly, we all recognize that soccer balls and bubbles have something in common: Both are concrete manifestations of the abstract idea of a sphere. These "simple" observations have been made by many people living in many different cultures at many different times in our history. Our ancestors may have begun to think about geometry and arithmetic almost as soon as they began to think at all.

Probability is different. Unlike geometry and arithmetic, whose origins lie in prehistoric times, the theory of probability is a comparatively recent discovery. We know when probability, as a branch of mathematics, was discovered. The theory of probability has its origins in the Renaissance with the work of Girolamo Cardano and later Galileo Galilei, but their work had little impact on those who lived later. The investigation of probability begins in earnest with an exchange of letters conducted by two French mathematicians,

Pierre de Fermat and Blaise Pascal. Mathematicians have been studying probability ever since.

One might think that the discovery of a new branch of mathematics means the solution of a new and difficult equation or the discovery of a new and exotic geometric shape. For probability, however, this is false. Many of the solutions to important problems early in the history of probability involved little more than simple arithmetic. This is not to say that just anyone could have solved these problems. These "simple" problems challenged some of the best minds of the day, because despite the elementary arithmetic involved in their solution, the concepts on which the solutions were based were new and challenging.

Despite the challenging nature of the subject, many mathematicians have found that time spent studying probability is time well spent. In fact, ever since its discovery less than 500 years ago probability has continued to attract the attention of some of the foremost mathematicians of each generation. Part of the fascination with probability can be traced to the interesting and surprising discoveries that have been made by using this branch of mathematics. Often these discoveries have shed new light on ordinary and familiar phenomena. For many of us that is what makes probability so interesting and challenging. Probability theory enables us to see and think about the world in new ways.

Although it is a relatively new branch of mathematics, probability theory is now one of the most used and useful of all the mathematical disciplines. It permeates our culture. One reason that probability has proved so useful is that it allows us to be very specific about the chance that some event will, or will not, occur. We tend to think of these kinds of events as random, but the adjective *random* often tells us more about ourselves than about the actual event.

To see this, it helps to look at a specific example. Consider, for example, the problem of forecasting the weather, an endeavor in which just about every prediction is expressed in the language of probability. When a meteorologist forecasts an 80 percent chance of rain tomorrow, that does not indicate that the meteorologist believes that weather is "random." Nor does it mean that there is no physical explanation for the rain, or that the laws of cause and effect do not apply to weather. It does mean that, given the current state of meteorological theory and the current set of measurements available from satellites and weather stations, the meteorologist still has some uncertainty about what the weather will be tomorrow. Presumably, as the physics of weather becomes better understood and better databases become available, the meteorologist's uncertainty about tomorrow's weather will continue to diminish and weather forecasting will continue to improve.

This leaves open the question of how we evaluate the meteorologist's prediction. For example, if the meteorologist predicts an 80 percent chance of rain and it does not rain, was the meteorologist wrong? Or more to the point: Was the meteorologist inaccurate? We can answer this question only if we know the meteorologist's record over the long run. On those days that follow a prediction of an 80 percent chance of rain, we should—if the meteorologist is accurate—find that it rained about 80 percent of the time. If, however, we discover that it rained 100 percent of the time or that it rained 60 percent of the time, we know that the 80 percent forecast is not a particularly accurate one. The theory of probability challenges us to think and learn in new ways.

With this example in mind, let us consider how probability can be used. Suppose we are trying to understand some process, which may be the weather or anything else. Ideally, we would like to predict the result of the process. That, however, is not always possible. We may not understand the process well enough to predict how it will turn out. Although we may not know enough to predict the precise result, we may, nevertheless, be able to list the *possible* outcomes of the process. To return to our meteorology example, we could list the possible outcomes as the elements of a set, such as {rain, no rain}. If we wanted to know more, we could make our list of possible outcomes more detailed. A more detailed weather list might consist of these possible outcomes: {rain with wind, rain without wind, no rain with wind, no rain without wind}. The list we make depends on what we know. What is important from the point of view of probability is that our list be a complete one—that is, if one event on the list does not occur, then some other event on the list must occur. One goal, then, is to determine the probability of each event on the list, but by itself accomplishing that goal is often not sufficient for many practical problems. It is often important to determine the factors on which our probability estimates depend because probabilities change. They reflect what we know. For example, what are the chances that it will rain tomorrow if we know that 100 miles west of our location it rained today? Information about probabilities does not allow us to predict with certainty what will happen next, but it does allow us to predict the frequency of some event or chain of events over the long run, and this ability can be very useful.

Ideas in the theory of probability are often subtle and even today are not widely appreciated. But although the ideas that underlie probability are somewhat obscure, the results of the theory of probability are used throughout our society. When engineers evaluate the safety of a nuclear reactor, they use probability theory to determine the likelihood that a particular component will fail and that the backup system or systems will fail as well. When engineers design a phone network they use probability theory to determine whether or not the network's capacity is large enough to handle the expected traffic. When health officials decide to recommend (or not to recommend) a vaccine for general use, their decision is based in part on a probabilistic analysis of the dangers that the vaccine poses to individuals as well as its value in ensuring the health of the general population. Probability theory plays an essential role in engineering design, safety analysis, and decision making throughout our culture.

The first part of this book traces the history of the theory of probability. We will look at some of the main ideas of the subject as well as the people who discovered them. We will also examine some applications of the theory that have proved important from the point of view of technology and public health. That latter part of the book is about statistics, which, as we will see, is the flip side of the theory of probability. We begin, however, with the history of some early "random event generators."

PART ONE

WHAT IS PROBABILITY?

1

THE IDEA OF RANDOMNESS

For most of us, the word *random* is part of our daily speech. We feel as if we know what *random* means, but the idea of randomness—random behavior, random phenomena, and random fluctuations—is an elusive one. How can we create random patterns? How can we recognize random patterns when we are confronted with them?

Central to the idea of randomness is the idea of unpredictability. A random pattern is often described as one that *cannot* be predicted. It is, however, difficult to build a theory—mathematical or otherwise—based on what something is not. Furthermore, this type of definition often tells us less about the pattern than it tells us about ourselves. A sequence of numbers may seem random, for example, but upon further study we may notice a pattern that would enable us to make better predictions of future numbers in the sequence: The pattern remains the same, but it is no longer as random as it first appeared. Does randomness lie simply in the eye of the beholder?

There is no generally agreed upon definition of randomness, but there have been several attempts to make explicit the mathematical nature of randomness. One of the best-known definitions of randomness is expressed in terms of random sequences of numbers. The precise definition is fairly technical, but the idea is not. To appreciate the idea behind the definition, imagine a very long sequence of numbers and imagine creating a computer program to describe the sequence. If every possible program that describes the sequence of numbers is at least as long as the sequence itself, then the sequence is random. Consider, for example, the sequence that

Knucklebone players. Games of chance have been a popular form of recreation for thousands of years. (© Copyright The British Museum)

begins with the number 0 and consists of alternating 0s and 1s {0, 1, 0, 1, 0, 1, . . .}. This sequence can be infinitely long, but it can be accurately described in only a few words. (In fact, we have already described it.) We conclude that the sequence is not random. Now suppose that we flip a coin and record the results so obtained in the following way: Each time we get "heads" we write the number 1, and each time we get "tails" we write the number 0. If we do this many times, we produce a very long sequence. The only way we can store the exact sequence—the precise series of 1s and 0s obtained by flipping the coin—is to store each of the numbers in the order in which it appeared. There is no way to compress all of the information about the sequence into a short description as we did for the sequence {0, 1, 0, 1, 0, 1, . . .}. Furthermore, a careful analysis of any part of the series will not—provided the coin is fair—enable us to predict future elements of the series with better than 50 percent accuracy. This sequence is random. Random sequences are incompressible.

Not every mathematician agrees with this definition of randomness; nor is it entirely satisfactory from a logical viewpoint. As the simpler definition given earlier did, it, too, defines randomness—or at least random sequences—in terms of what they are not: They are not compressible. There are, nevertheless, some positive characteristics of this more mathematical definition. Part of this definition's attraction lies in the fact that it enables researchers to investigate degrees of randomness. If a sequence can be partly compressed, then it is less random than it first seemed. If this more modern definition is not the best one possible, it is, at least, a step in the right direction.

Although the notion of randomness is difficult to define, it is, nevertheless, an idea that has made its way into our daily lives in a variety of ways. Most modern board games, for example, incorporate some aspect of randomness. Often this involves rolling one or more dice. Dice are a common choice for a *randomizing agent*, a device used to produce a random sequence or pattern, because the patterns obtained by rolling dice are stable enough to make the overall flow of the game predictable: We do not know which number will appear on the next roll of the dice, but we do know that over the long run all numbers will appear with predictable frequencies. This type of stability makes it possible to plan game strategy rationally.

The other application of random processes that is of special interest to us is the use of random processes as an aid in decision making. Athletic teams, for example, use a random process as an aid in decision making whenever they toss a coin to determine which team takes possession of the ball first. Other, similar uses are also common. For example, in choosing between two alternatives, such as whether to go to the movies or the park, we may well use a coin: "Heads we go to the movie; tails we go to the park." Flipping a coin is often perceived as a method to decide impartially between two competing alternatives. On a more sophisticated level, computer programs sometimes incorporate a random number generator—a secondary program designed to choose a number "at random" from some predetermined set—into the main program so that the computer can perform certain calculations without introducing bias. "Fairness" is key: Coins, dice, cards, and random number generators are usually perceived as devices that generate numbers unpredictably and without bias.

The incorporation of randomness into recreational activities and decision-making processes is not new, of course, but in many ways the interpretations and expectations that we have about the processes are. There is ample evidence that the earliest of civilizations used random processes in just the same way that we do today, but their expectations were quite different from ours. In fact, in many cases, they simultaneously used random processes even as they denied the existence of randomness.

Randomness before the Theory of Probability

How old is the search for random patterns? Archaeologists have found prehistoric artifacts that appear as if they could have been used in the same way that we use dice today. Bits of bone and carefully marked stones that have been unearthed at prehistoric sites were clearly created or at least put aside for a purpose. These objects evidently had meaning to the user, and they resemble objects that were later used in board games by, for example, the ancient Egyptians. This evidence is, however, difficult to interpret. Without a written record it is difficult to know what the artifacts meant to the user.

One of the earliest devices for producing random patterns for which there is direct evidence is the astragalus, a bone found in the heels of deer, sheep, dogs, and other mammals. When thrown, the astragalus can land on any of four easy-to-distinguish sides. Many astragali have been found at prehistoric sites, and it is certain that they were used in ancient Egypt 5,000 years ago in games of chance. There are pictures of Egyptians throwing astragali while playing board games. Unfortunately, the only record of these early games is a pictorial one. We do not know how the game was played or how the patterns produced by throwing the astragali were interpreted.

The earliest game of chance that we understand well enough to play ourselves is one from Mesopotamia. The Mesopotamian civilization was located inside what is now Iraq. It was one of the oldest, perhaps *the* oldest, literate civilization in history. The earliest written records we have from this culture are about 5,000 years old. Babylon was the most famous city in Mesopotamia, and another important city was Ur. While excavating graves at Ur during the early 20th century archaeologists uncovered a board game that had been buried with its user. The board game, which is beautifully crafted, is about 4,500 years old. We can be sure that it is a board game—we even know the rules—because ancient references to the game have also been unearthed. This game is called the Game of 20 Squares. It is played by two people, each of whom relies on a combination of luck and a little strategy to win. The luck part involves rolling dice, to determine how many squares each player can move his or her piece. The skill part involves choosing which piece to move. (You can play this most ancient of all known board games on the website maintained by the British

The Game of 20 Squares was played for 3,000 years—until the first millennium of the common era. This particular board dates from about 2500 B.C.E. (© Copyright The British Museum)

Museum at http://www.mesopotamia.co.uk/tombs/challenge/ ch_set.html. They call it the Royal Game of Ur.) What is important to us is that the game develops in a more or less random way, because the number of spaces each player can jump is determined by a throw of a set of dice.

The Game of 20 Squares was played for millennia over a wide area of the world, including Egypt and India as well as Mesopotamia. It was one of the most successful board games of all time, but it did not inspire a theory of probability. There is no indication that anyone tried to devise an optimal strategy for winning the game based on the probability of certain outcomes of the dice.

Two thousand five hundred years after the invention of the Game of 20 Squares, Mesopotamian culture was on the wane. The dominant culture in the area was Rome, and the inhabitants of ancient Rome loved to gamble. Gambling, or gaming, can be described as the board game minus the board. Skill is eliminated as a factor, and participants simply bet on the outcome of the throw.

Gambling, then as now, however, was associated with many social problems, and the Romans had strict laws that limited gambling to certain holidays. These laws were widely ignored, and the emperors were some of the worst offenders. The emperors Augustus (63 B.C.E.–A.D. 14) and Vitellius (A.D. 15–69) were well known as inveterate gamblers. They enjoyed watching the random patterns emerging as they threw their astragali again and again—astragali were more popular than dice as devices for creating random patterns—and they enjoyed cheering when the patterns went their way.

The rules of the games were simple enough. A popular game involved "throwing" several astragali. When a player threw an unlucky pattern he or she placed money into the pot. The pattern continued with each player's adding money to the pot until a player threw a "lucky" combination of astragali; then she or he won all of the money in the pot, and afterward the game began again. It does not appear that the Romans were interested in thinking about randomness on a deeper level, although they had plenty of opportunities to do so. In the following excerpt of a letter that Emperor Augustus sent to one of his friends he describes how he spent the day of a festival:

> We spent the Quinquatria very merrily, my dear Tiberius, for we played all day long and kept the gaming-board warm. Your brother made a great outcry about his luck, but after all did not come out far behind in the long run; for after losing heavily, he unexpectedly and little by little got back a good deal. For my part, I lost twenty thousand sesterces, but because I was extravagantly generous in my play, as usual. If I had demanded of everyone the stakes which I let go, or had kept all that I gave away, I could have won fully fifty thousand. But I like that better, for my generosity will exalt me to immortal glory.
>
> *(Suetonius, Suetonius, trans. J. C. Rolfe [Cambridge, Mass.: Harvard University Press, 1913])*

This is clearly a letter from someone who expects nothing more from gambling than a good time and immortal glory. This attitude was typical of the times.

In ancient times astragali, dice, the drawing of lots, and other randomizing agents were also used as aids in decision making. A list of possible actions was drawn up and each action assigned a number or pattern; then the dice or astragali were thrown and the outcome noted. The chosen course of action was determined by the pattern that appeared. This type of decision making was often associated with religious practice, because the participants saw the outcome as an expression of providence. By using what we might call a randomizing agent the questioner had released control of the situation and turned over the decision to his or her god, an interpretation of a mode of decision making that is not restricted to antiquity. Today there are many people who continue to hold that what are often described as random actions are actually expressions of divine will.

Although there are many instances in antiquity of interpreting a random outcome as the will of God, there is no more articulate expression of this idea than a legal opinion written in the highly

The Supreme Court justice Henry Baldwin (Painted by Thomas Sully, *Collection of the Supreme Court of the United States*)

publicized 19th-century criminal trial *U.S. v. Holmes* (1842). The judge who wrote the opinion was a Supreme Court justice, Henry Baldwin. He was sitting in for a Philadelphia trial judge when he heard this case. Here are the facts:

A ship, the *William Brown*, was carrying 80 passengers across the North Atlantic when it struck an iceberg. There were two boats aboard the ship that could be used as lifeboats. One boat was much smaller than the other. The *William Brown* sank with 30 passengers, mostly children, aboard. After some initial shuffling, the small boat, which was outfitted with oars and a sail, carried eight passengers including the captain. The larger boat, which was only 22 feet long, carried 42 passengers including a few crew members and the mate. The larger boat was severely overloaded, leaking, and sitting very low in the water. The passengers had to bail steadily to prevent it from sinking. It did not have a sail and, in any case, was too heavily loaded to do anything except drift. The smaller boat sailed for Canada, where it was rescued by a Canadian fishing vessel. After the larger boat had drifted for about a day on the open sea, the wind picked up. Waves swamped it even though the passengers bailed frantically. The mate ordered the crew to lighten the boat. Two sailors threw some of the passengers overboard, and they soon drowned. In this way the crew raised the level of the boat enough that it could ride the waves. This action saved the crew and the remaining passengers. The boat drifted eastward and was eventually rescued by a French ship and taken to a French port.

Later, when the survivors reached Philadelphia, they spoke in favor of prosecuting the sailors for murder. It was his misfortune that Holmes, who was involved in throwing the passengers overboard, was the only sailor whom the authorities could locate. The grand jury refused to indict him for murder so he was indicted for voluntary manslaughter. After much ado Holmes was sentenced to six months in jail and a $20 fine. (He served the jail sentence but did not pay the fine because he received a pardon from President John Tyler.) Explaining the court's decision, the presiding judge, Supreme Court justice Henry Baldwin, wrote, in part

> there should be consultation, and some mode of selection fixed, by which those in equal relation may have equal chance for their life . . . when a sacrifice of one person is necessary to appease the hunger of others, the selection is by lot. This mode is resorted to as the fairest mode, and in some sort, *as an appeal to God*, for selection of the victim.

The emphasis is ours. It was Justice Baldwin's thinking that the sailors, except those whose navigation duties made them indispensable, should have been at the same risk of being thrown overboard as the passengers. Their mistake, he believed, lay in putting themselves above the passengers. The sailors, as the passengers were, should have been subject to a chance procedure whose outcome would determine who would be thrown overboard. In the statement cited we can see how Justice Baldwin sees randomness as an opportunity for a deity to intercede.

Early Difficulties in Developing a Theory of Randomness

It is apparent that randomizing agents were an important part of ancient societies just as they are of today's. Despite this, ancient societies did not develop a theory of randomness. There was nothing in any ancient society that corresponded to the theory of probability. This is not because ancient peoples were not mathematically sophisticated. Many of them were. Furthermore,

RANDOMNESS AND RELIGION TODAY
IN BURKINA FASO

Today, in the country of Burkina Faso, which is located in western Africa, lives a group of people called the Lobi. (Burkina Faso means "land of the honest people.") Traditional Lobi beliefs hold that some men and a few women can communicate with mystical beings called *thila*. These people are "diviners." The Lobi consult the thila about a wide variety of topics, but communicating with the thila can take place only with the help of a diviner. The role of the diviner in Lobi society is very interesting and in some ways inspiring, but from our point of view it is the method with which the diviner communicates with the thila that is of interest. At a certain point in the ceremony the diviner asks questions of a particular thila, so that the diviner can be sure that he or she has divined correctly. We can, if we so choose, understand the verification procedure of the diviner in terms of random patterns. The diviner uses cowry shells to form a random pattern. Cowry shells have one flat, open side and one curved, closed side, so a cowry shell can land either flat side up or curved side up. There are no other possibilities. The diviner rolls two or more cowry shells. If one shell lands flat side up and all other shells land curved side up, this pattern is interpreted as a positive answer by the thila. A no from the thila is understood if any other configuration of cowry shells is rolled. This is a nice example of how what we might perceive as a random pattern is interpreted by others as not random at all. The randomness is, instead, an opportunity for a deity to communicate directly with the diviner.

many of the early problems in the theory of probability were not mathematically difficult; they were well within the range of mathematicians living in China, India, Mesopotamia, Greece, and several other places. Why, then, was the development of the theory of probability delayed until the 16th century?

The first barrier to progress in developing a theory of randomness was essentially technical. In antiquity, the principal randomizing agent was often the astragalus, and the structure of astragali are decidedly not uniform. An astragalus has an irregular shape. More importantly, the shape and weight distribution of an astragalus depend very much on the age and species of the animal from which it was obtained. Consequently, the frequency of various outcomes depends on the particular astragali used. Changing astragali in the middle of a game amounts to changing the game, because the change also alters the frequency of various outcomes. It is not possible to develop uniform data (or a uniform theory) for astragali in the same way that one can for modern dice. The fact that astragali were not uniform probably inhibited the development of a theory of randomness based on their use. It certainly would have limited the usefulness of such a theory. (It is also worth noting that what has been said of astragali can also be said of many early dice. These often were not exactly cubical; nor did they always have a uniform weight distribution. No one would use such asymmetric dice today, but at one time they were common.)

In contrast to these early randomizing agents, modern dice are uniform in structure: A well-made die is a cube; its weight is distributed evenly throughout, and as a consequence every such die is as likely to land on one side as on another. This is the so-called fair die. Over the long run the frequencies of all the outcomes obtained by rolling any such die are the same. This type of stability makes it possible to compare a single set of theoretical predictions of frequencies with empirical data obtained from *any* die because what is true for one modern die is true for them all. The existence of good approximations to the ideal "fair" die made a big difference. Good approximations provided an accurate physical representation of an ideal concept. As well-made dice replaced the astragali, and as well-made cards became more affordable, it became possible to develop a theory of randomness based on "fair," well-understood randomizing agents. Furthermore, there was great interest among gamblers and others in such a theory for its possible utility.

A second, more fundamental barrier to the development of probability was the difference between ancient and modern perceptions about the use of random processes as an aid in decision making. As pointed out in the first section of this chapter, when we flip a coin to decide between two alternatives, we are often appealing to a random and unbiased process. We are simply looking for a means to distinguish between competing alternatives when neither alternative is favored. It may seem that the use of randomizing agents by the ancients—and the type of selection process favored by Justice Baldwin in *U.S. v. Holmes*—is similar to the more modern conception of such agents, but that similarity is only superficial. If one perceives that random outcomes are actually expressions of divine will, then one does not truly believe that the actions are random at all. This is a more profound barrier to the development of a theory of probability than the technical differences between uniform dice and nonuniform astragali, because it is a conceptual barrier. With the older understanding of random events as expressions of divine will there is no need to search for stable frequencies; they have no meaning. No matter what past data indicate, future frequencies can always change, because every outcome is the reflection of conscious decisions made by an intelligent being.

The idea that a random process is not random but instead subject to manipulation by God or even the "skilled" has proved to be a very tenacious one. It was not until mathematicians began to reject the ideas of divine intercession and of luck—and the rejection was very tentative at first—that the theory of probability began to develop. The shift toward a new type of reasoning—a new *perception* of unpredictable phenomena—began to occur in 16th-century Italy with the work of Girolamo Cardano.

2

THE NATURE OF CHANCE

The Italian mathematician Girolamo Cardano (1501–76), also known as Jerome Cardan, was the first to write in a somewhat modern way about the odds of throwing dice. His interest in rolling dice is understandable. He loved to gamble. He loved to play chess and bet on the outcome. He was also a prominent physician as well as a mathematician, astrologer, and scientist. He lived in Italy during the Renaissance and contributed to knowledge in a variety of fields. Cardano was a Renaissance man—smart, self-confident, and self-absorbed. He wrote at length about himself, and he enjoyed describing and praising his own accomplishments.

Girolamo Cardano, the first mathematician to attempt to formulate a mathematical theory of probability (Library of Congress, Prints and Photographs Division)

(In retrospect, it is clear that he sometimes claimed credit for ideas and accomplishments that were not entirely his own.)

Things did not come easily to Girolamo Cardano. He wanted to be admitted to the College of Physicians in Milan but was refused twice. He succeeded on his third attempt. The process of gaining admission to the college took years, but Cardano was not someone who became easily discouraged. He believed in himself, and with good reason. He eventually became a well-known and much-sought-after physician. Today Cardano is best remembered as a mathematician and the author of the book *Ars Magna*, a book about algebra that is still in print more than 400 years after it was first published. Some claim that Cardano's book was the start of the modern era in mathematics. It certainly made a big impression on

A standard die, flattened to show the relative positions of the numbers

his contemporaries. Cardano, however, wrote many books on many different subjects, including chess and dice, two games in which he seems to have lost a lot of his money. That he had a gambling problem is clear. In one story he proudly recounts how he was able to recoup his losses: "Thus the result was that within twenty plays I regained my clothes, the rings, and a collar for the boy" (Ore, Oystein, *Cardano, The Gambling Scholar*, Princeton, N.J.: Princeton University Press, 1953. Used with permission). Some of what he wrote about chance, in particular, was not new even for his time, but there are places in his book *Liber de Ludo Aleae* where we can find the barest beginning of the idea of probability.

Girolamo Cardano expended a great deal of energy thinking about games of chance. In *Liber de Ludo Aleae* he writes about dice, cards, astragali, and backgammon. It was not an easy subject for him. He was beginning to think about an old problem in a new way. When Cardano wrote about a single die he clearly had an ideal, or fair, die in mind. His writings on the subject are not entirely clear, however, and there are many issues—for example,

CARDANO'S MISTAKE

Cardano asserted that if one throws a die three times the chance that a given number will show at least once is 50 percent. This is now recognized as the wrong answer. To understand the right answer, one needs to know three facts about probability.

1. Each roll of a die is *independent* of every other roll. Here *independent* has a technical meaning: No matter what the outcome of any past roll—or any series of past rolls—the probability of every future outcome remains unchanged.
2. The probability that a given event will occur plus the probability that the event will not occur always adds up to 1. In symbols, if *p* is the probability that some event will occur, then the probability that this event will not occur is always 1 − *p*.
3. When two events are independent the probability that both will occur is the product of their individual probabilities. Consider, for example, two events, which we will call *A* and *B*. If the probability that *A* will occur is *p* and the probability that *B* will occur is *q*, then the probability that *A* and *B* will occur is *p* × *q*.

To compute the probability that a number will show at least once in three throws of a die, it is easier to compute the probability that the number will fail to show even once and subtract this probability from 1. (See fact 2.) The probability that the number will not show on a single throw of the die is 5/6.

By the first fact, each throw is independent, so the probability that the number will not appear on the second throw is also 5/6. The same is true of the third throw. By the third fact, the probability that the number will fail to appear on all three throws is 5/6 × 5/6 × 5/6 or 125/216 or approximately 58 percent. By fact 2, the probability that the given number will appear at least once is 1 − 0.58 or 42 percent.

the odds of rolling a particular sequence of numbers—that he does not address. Nevertheless, he clearly saw that the pastime he loved had some mathematical basis, because he mathematically compared the odds of various simple outcomes.

CARDANO ON LUCK AND MATH

Girolamo Cardano is a transitional person in the history of probability. Of course, every mathematician worthy of note is, in some sense, a transitional figure; each good mathematician corrects past errors and contributes something to future progress. But the statement has a special meaning for Cardano. Because of his mathematical background he was able to identify a new way of thinking about games of chance.

Cardano was sometimes able to understand and use probability in ways that sound modern. For example, he knew that the odds of throwing a 10 with two dice are 1/12. He finds this by counting the number of favorable outcomes. There are, he tells us, three ways of obtaining a 10 with two dice. One can roll

- (5, 5), that is, a 5 on each die, or

- (6, 4), that is, a 6 on the first die and a 4 on the second, or

- (4, 6), a 4 on the first die and a 6 on the second.

Next notice that there are 36 different outcomes. To see why, imagine that one red and one green die are used—that way we can distinguish between them. If 1 is rolled with the red die, that 1 can be paired with any of six numbers—that is, 1, 2, 3, 4, 5, and 6—rolled on the green die. So there are six possible outcomes associated with rolling a red 1. There is, however, nothing special about the number 1. Exactly the same argument can be used for any other number that appears on the red die. Summing up all the possibilities we get 36 different possible outcomes. (See the accompanying chart.)

Divide the sum of favorable outcomes (3) by the number of possible outcomes (36) and one obtains 3/36 or 1/12. It is a simple result, but it shows that he understands the principle involved.

What is interesting about Cardano is that although he understands how to calculate the odds for certain simple outcomes, he does not quite believe in the calculation. The difficulty that he has in interpreting his calculations arises from the fact that he cannot quite jettison the very unscientific idea of luck. Here is an excerpt from a section of *Liber de Ludo Alea* entitled "On Timidity in the Throw."

For this reason it is natural to wonder why those who throw the dice timidly are defeated. Does the mind itself have a presentiment of evil? But we must free men from error; for although this might be thought true, still we have a more manifest reason. For when anyone begins to succumb to adverse fortune, he is very often accustomed to throw the dice timidly; but if the adverse fortune persists, it will necessarily fall unfavorably. Then, since he threw it timidly, people think that it fell unfavorably for that very reason; but this is not so. It is because fortune is adverse that the die falls unfavorably, and because the die falls unfavorably he loses, and because he loses he throws the die timidly.

In *Liber de Ludo Aleae* we find luck and math side by side. That is part of what makes the book so interesting to a modern reader.

"First" Die						
+	**1**	**2**	**3**	**4**	**5**	**6**
1	2	3	4	5	6	7
2	3	4	5	6	7	8
3	4	5	6	7	8	9
4	5	6	7	8	9	10
5	6	7	8	9	10	11
6	7	8	9	10	11	12

(left axis label: Second Die)

The table shows all 36 possible outcomes that can be obtained by rolling two dice. The three shaded squares indicate the three possible ways of rolling a 10:6 on the first die and 4 on the second, 5 on each die, and 6 on the second die and 4 on the first.

A modern reader can occasionally find it a little frustrating (or a little humorous) to read the *Liber de Ludo Aleae*. One begins to wonder when Cardano will get around to drawing the "obvious" conclusions. He usually does not. He points out, for example, that if one chooses any three sides of a die, then the numbers on those three sides are just as likely to show on one roll of the die as the numbers on the other three sides. From this he concludes, "I can as easily throw one, three or five as two, four or six" (ibid.). In a sense, by marking out three faces of a six-sided die as favorable and three as unfavorable he turned the problem of rolling a die into a coin-toss problem: The odds are 50/50, he tells us, that we will roll either a 1, a 3, or a 5. He was right, of course, and he did go a little beyond this simple case, but his understanding of probability, even as it relates exclusively to dice, was very limited.

Mathematically, he came very close to making deeper discoveries, but he never quite made the necessary connections. Moreover, not every mathematically formulated remark that he wrote about dice is correct. He concludes, for example, that if one throws a die three times the chance that a given number will show at least once is 50 percent, whereas it is actually about 42 percent. To be sure, he did not get very far in his analysis, but it is important to keep in mind that he was the first to attempt to formulate probabilistic descriptions of "random phenomenon."

We can develop a fuller appreciation of Cardano's work if we keep in mind two additional barriers that Cardano faced in addition to the newness of the subject. First, it would have been hard for anyone to develop a more comprehensive theory of probability without a good system of algebraic notation. Without algebra it is much harder to represent one's mathematical ideas on paper, and in Cardano's time the algebraic notation necessary for expressing basic probability was still in the process of being developed. (*Liber de Ludo Aleae* is practically all prose.) Second, although Cardano stood at the edge of a new way of thinking about randomness, it is clear that he could not quite let go of the old ideas. In particular, he could not lose the old preconceptions about the role of luck. He was very sure, for example, that the attitude of the person throwing the dice affects the outcome of the throw. (Over a century later the great mathematician Abraham de Moivre felt it necessary to include a section in his book *The Doctrine of Chances* repudiating the idea that luck is something that can affect the outcome of a random event.) Although he could compute simple odds, Cardano was unwilling to let the numbers speak for themselves. Luck, he believed, still played a part.

Despite these shortcomings we find in Cardano's writings the first evidence of someone's attempting to develop a mathematical description of random patterns.

Galileo Galilei

The Italian scientist Galileo Galilei (1565–1642) was one of the most successful scientists of all time. His astronomical observations, especially of Venus, the Sun, and the planet Jupiter, provided powerful proof that the Earth is not at the center of the universe. He was one of the first scientists to investigate physics using a combination of carefully designed experiments and painstaking mathematical analysis. He played an important role in establishing the foundations of modern science. He demonstrated creativity in the pursuit of scientific and mathematical truth and bravery in the face of adversity. In his article "Thoughts about Dice-Games," he also wrote a little about randomness.

Galileo's observations on dice are not well known. Even Galileo did not seem to pay much attention to the subject. He states in the first paragraph that he is writing about dice only because he was "ordered" to do so. (He does not mention who ordered him.) Galileo seems to have been the only person of his time thinking about randomness in a mathematical way. (Cardano died when Galileo was a boy.) The ideas that Galileo expresses in his paper are simply and directly stated. Even today this very short paper makes a nice introduction to the simplest ideas about probability.

Galileo is especially interested in the problem of explaining why the numbers 10 and 11 appear more frequently in throws of three dice than do the numbers 9 and 12. The solution is simply a matter of counting. He begins by noting that there are only 16 different numbers that can be obtained by rolling three dice: 3, 4, 5, . . ., 18. These numbers are not all equally likely, however. The number 3,

Galileo Galilei. Although he spent little time thinking about probability, he saw more deeply into the subject than anyone before him. (Library of Congress, Prints and Photographs Division)

he notes, can be obtained in only one way: three 1s must be rolled. Other numbers are more likely to appear than 3 because they can be obtained by a greater variety of combinations of the dice.

To determine why 10 and 11 are more likely numbers than 9 and 12 when rolling three dice, Galileo counts all of the ways that the numbers 10 and 11 can be obtained. He shows, for example, that there are 27 different ways of rolling a 10 but only 25 different ways of rolling a 9. To see why this is true, imagine that the three dice are identical in every way except color. Suppose that one die is green, the second yellow, and the third red. Now that we can easily distinguish the dice, we can see that two green, one yellow, one red is a different outcome from one green, two yellow, one red. This is true even though in both instances the dice add up to 4. With this in mind it is a simple matter of counting all possibilities. The accompanying table lists all possible combinations of 9 and 10 for comparison.

Notice, too, that there are 216 different possible outcomes associated with rolling three dice: six different "green" numbers, six yellow, and six red. Since the numbers on the differently colored dice can occur in any combination the total number of combinations is $6 \times 6 \times 6$ or 216 outcomes.

If we were to study Galileo's problem ourselves we would probably conclude our study with the observation that the chances of rolling a 10 are 27/216, because there are 27 different ways of rolling a 10 out of a total of 216 distinct possible outcomes. By contrast, the chances of rolling a 9 are 25/216. Galileo does not go this far. He is content to list the total number of outcomes that yield a 10 (27 combinations) and the total number of outcomes that add up to 9 (25 combinations) and then conclude that a 10 is more likely than a 9. Galileo does not use any of the language that we would associate with probability or randomness. To him it is a simple matter of counting and comparing. Nevertheless, Galileo's

COMBINATIONS OF THREE DICE THAT SUM TO 9 (25 SUCH COMBINATIONS)	COMBINATIONS OF THREE DICE THAT SUM TO 10 (27 SUCH COMBINATIONS)
(1, 2, 6)	(1, 3, 6)
(1, 3, 5)	(1, 4, 5)
(1, 4, 4)	(1, 5, 4)
(1, 5, 3)	(1, 6, 3)
(1, 6, 2)	(2, 2, 6)
(2, 1, 6)	(2, 3, 5)
(2, 2, 5)	(2, 4, 4)
(2, 3, 4)	(2, 5, 3)
(2, 4, 3)	(2, 6, 2)
(2, 5, 2)	(3, 1, 6)
(2, 6, 1)	(3, 2, 5)
(3, 1, 5)	(3, 3, 4)
(3, 2, 4)	(3, 4, 3)
(3, 3, 3)	(3, 5, 2)
(3, 4, 2)	(3, 6, 1)
(3, 5, 1)	(4, 1, 5)
(4, 1, 4)	(4, 2, 4)
(4, 2, 3)	(4, 3, 3)
(4, 3, 2)	(4, 4, 2)
(4, 4, 1)	(4, 5, 1)
(5, 1, 3)	(5, 1, 4)
(5, 2, 2)	(5, 2, 3)
(5, 3, 1)	(5, 3, 2)
(6, 1, 2)	(5, 4, 1)
(6, 2, 1)	(6, 1, 3)
	(6, 2, 2)
	(6, 3, 1)

paper is the most advanced treatise on a problem that we would treat with the mathematics of probability that had been written up until that time. Perhaps even more importantly, it is free of the idea of luck—a concept that had marred Cardano's thinking. It was an important accomplishment despite the fact that no one, apparently not even Galileo himself, considered it worthy of much attention.

Pierre de Fermat and Blaise Pascal

The theory of probability is often said to have begun with the work of two Frenchmen, Blaise Pascal (1623–62) and Pierre de Fermat (1601–65). They were both extremely successful mathematicians. Each of them made many discoveries in a variety of mathematical disciplines, but neither Fermat nor Pascal was primarily a mathematician. Both were mathematical hobbyists; fortunately, they were brilliant hobbyists.

Pierre de Fermat was 22 years older than Pascal. He studied law at the University of Toulouse and later found work with the government in the city of Toulouse. This allowed him to work as a lawyer and to pursue the many interests that he had outside the law. When the law courts were in session he was busy with the practice of law. When the courts were out of session he studied mathematics, literature, and languages. Fermat knew many languages, among them Greek, Latin, Spanish, Italian, and, of course, French. He was well liked. By all accounts Fermat was polite and considerate and well educated, but beneath his genteel exterior he was passionately curious.

Mathematics is a difficult subject to pursue in isolation. The ideas involved can be conceptually difficult, and the solutions can be technically difficult. It is easy to get bogged down with details and miss the forest for the trees. To keep one's mind fresh it helps to have access to other people with similar interests. For Fermat, "keeping fresh" meant sending letters to accomplished mathematicians. He maintained a lively correspondence with many of the best mathematicians of his time. The letters, many of which were preserved, show a modest and inquisitive man in a serious and sustained search for mathematical truth.

In contrast to Fermat, Blaise Pascal spent his teenage years gleaning his mathematical education from face-to-face contact with some of the finest mathematicians in Europe. He accomplished this by attending one of the most famous math "clubs" in the history of the subject.

In France and Italy during the time of Fermat and Pascal, and even during the time of Cardano, there existed many formal and informal groups of like-minded individuals who met together to discuss new ideas in science and mathe-

Ancient dice and a shaker made of bone (Museum of London/Topham-HIP/The Image Works)

matics. Meetings were held more or less regularly. One of the most famous of these groups met each week in Paris, Pascal's hometown, at the house of Marin Marsenne. Marsenne was a priest with a love of science, mathematics, and music. He was a prolific writer and corresponded with many of the leading mathematicians and scientists of his day, but it was the meetings, held weekly at his house, that made him well known throughout Europe. Some of the finest mathematicians and scientists of the time spent one evening each week at what came to be known as the Marsenne Academy. They talked, they argued, and they learned. Pierre de Fermat, who lived in far-away Toulouse, was not a member, but another mathematician, Etienne Pascal, was frequently in attendance. In addition to his attendance at the academy, he and Fermat corresponded on a number of subjects. Although Etienne Pascal was a good mathematician, he is best remembered today as the father of Blaise Pascal.

Etienne Pascal, as did Fermat, worked as a civil servant, but his principal interest was his son's education. Initially, he instructed Blaise in languages and literature. He would not teach him

Blaise Pascal. His brief exchange of letters with Pierre de Fermat opened up a new way of thinking about random processes. (Salaber/The Image Works)

mathematics, because he did not want to overwork his son. It was not until the younger Pascal began to study geometry on his own that his father relented and began to teach him math as well. Blaise Pascal was 12 when he began to receive instruction in mathematics. By the time he was 14 years of age he was accompanying his father to the get-togethers at Father Marsenne's house.

The meetings had a profound effect on Blaise Pascal's thinking. By the time he was 16 he had made an important discovery in the new field of projective geometry. (The mathematician who founded the field of projective geometry, Gérard (or Girard) Desargues, attended the meetings regularly, and Pascal's discovery was an extension of the work of Desargues.) The younger Pascal's interests changed quickly, however, and he soon stopped studying geometry. By the time he was 18 he was drawing attention to himself as the inventor of a mechanical calculator, which he created to help his father perform calculations in his capacity as a government official. The Pascaline, as it came to be called, was neither reliable nor cheap, but he made several copies and sold some of them. These calculators made a great impression on Pascal's contemporaries, and several later calculators incorporated a number of Pascal's ideas into their design.

As an adult Pascal was acquainted with a French nobleman, the chevalier de Méré, a man who loved to gamble. Pascal and de Méré discussed the mathematical basis for certain problems associated with gambling. Pascal eventually turned to Fermat for help in the solution of these problems. In 1654, Fermat and Pascal began a famous series of letters about games of chance.

Some of the problems that Pascal and Fermat discussed concerned "the division of stakes" problem. The idea is simple enough. Suppose that two players place equal bets on a game of chance. Suppose that one player pulls ahead of the other and then they decide to stop the game before it has reached its conclusion. How should they divide the stakes? If one player is ahead then it is unreasonable to divide the stakes in half since the player who is ahead would "probably" have won. As every gambler knows, however, being ahead in a game of chance is no guarantee of a win: In fact, sometimes the player who is behind eventually wins anyway. Nevertheless, over the long run the player who is ahead wins more often than the player who is behind. The division of the stakes should reflect this. This problem involves several important probability concepts and may have been inspired by ideas outside the field of gambling. (See the sidebar.)

In their letters Pascal and Fermat solve multiple versions of this type of gambling problem. They began with problems that involve two players and a single die. Later, they considered three-player games, but they did not limit themselves to the division of stakes problem. They also answered questions about the odds of rolling a particular number at least once in a given number of rolls. (What, for example, are the odds of rolling a 6 at least once in eight rolls of a die? See the sidebar Cardano's Mistake earlier in this chapter for the solution to a closely related problem.) Their letters reflect a real excitement about what they were doing.

Unfortunately, Pascal and Fermat corresponded for only several months about games of chance, and then Pascal stopped working in mathematics altogether. He joined a religious order and gave up mathematics for the rest of his life. Several years later, Fermat sent Pascal one final letter offering to meet him halfway between their homes to visit, but Pascal refused. In a few more years both men were dead.

The sophistication of Fermat and Pascal's work far surpassed that of the work of Cardano and Galileo. Previously, Cardano had asserted that what he had discovered about a single die was interesting from a theoretical viewpoint but was worthless from a practical point of view. It is true that neither his discoveries nor any subsequent discoveries enable a gambler to predict which

THE DIVISION OF STAKES, AN ALTERNATIVE INTERPRETATION

One of the most important problems in early probability theory was called the division of stakes. The problem was often described in the following terms:

Two players agree to a game of chance. They wager equal amounts of money on the outcome. All money goes to the winner. The game begins but is interrupted before it is completed. One player is ahead when the game ends. How should the stakes be divided?

In the main body of the text this is described as a problem that was motivated by gambling concerns, but there is another interpretation that is of interest to us here. Some scholars believe that the division of stakes problem was motivated by broader economic concerns. During the Renaissance, lenders and merchants began to develop more sophisticated systems of finance. Lenders sought to loan merchants money for their businesses in the hope that the merchants would return to them the capital plus an additional sum (the lender's profit) at a future date. (Today we often think of the profit as interest charged on the loan, but there were other, alternative strategies in practice at the time such as a share of the merchant's future profits.) Merchants were expected to risk their own money on the venture as well, so that the risk was shared.

The question then arose as to what were fair terms for the risk assumed by each party: In the event that the situation did not develop as the lender and merchant anticipated how could the "stakes" be fairly divided between them? Seen in this way, the gambling questions to which these early theorists addressed themselves—the questions on which the theory of probability was originally founded—were really problems in insurance stated in terms of recreational gambling. This would also help to explain why these types of gambling problems developed when they did. Europe's economy underwent a period of rapid change and growth at the same time that mathematicians became interested in the division of stakes problem. Some scholars believe that the two phenomena were related.

number will turn up on the next roll of a die; by their nature random processes are unpredictable. (If they were predictable they would not be "random.") What Fermat and Pascal discovered instead was that they could (in some simple cases, at least) predict properties of the random pattern that would emerge if the dice were rolled many times. For example, although they could not determine whether or not a gambler would roll a 6 at least once in eight rolls of a single die—because they *could not* predict individual events—they *could* predict how frequently the gambler would roll at least one 6 in eight rolls of a single die *if the gambler performed this "experiment" many times.* This type of insight, which allows one to compare the likelihood of various outcomes, can be useful from a practical point of view. Over the course of their brief correspondence they made a serious effort to apply the results of the new mathematics to problems in gaming, and in the process they discovered a new way of thinking about randomness.

We should be careful not to overstate what Fermat and Pascal did. They solved a set of isolated problems in probability; they did not develop a broad theory. This is not surprising given the brief time that they worked on these problems. When putting their accomplishments into perspective, it helps to compare their results with Euclidean geometry, a subject with which they were both very familiar. In Euclidean geometry Greek mathematicians had identified the objects with which they were concerned, points, lines, planes, and the like. They made a list of definitions and axioms that described the basic properties of these objects. Finally, they used these fundamental properties to deduce still other properties of the system of points, lines, and planes that they had imagined into existence. Greek mathematicians attempted to create a complete mathematical system. They wanted to create a purely deductive science. Pascal and Fermat's work was not on this level. In fact, mathematicians would not take a deep look into the ideas underlying the theory of probability until the 20th century.

Nevertheless, the letters that Pascal and Fermat exchanged made a strong impression on many mathematicians. At first, their discoveries just heightened interest in the mathematical theory of gambling, but these kinds of results were soon used in surprising and important ways. Random patterns were soon used in everything from the computation of the number π to the establishment of rational public health policy. In a very real sense the history of probability begins with Pascal and Fermat.

Christian Huygens

The short-term effect of the work of Pascal and Fermat was to inspire discussion among many of the mathematicians in Paris. One of those to hear and take part in these discussions was a young Dutch mathematician, Christian Huygens (1629–95). As is Galileo, Christian Huygens is now remembered primarily as a physicist and inventor. He developed a new telescope design and was the first to understand the nature of Saturn's rings. (Galileo's telescope produced blurry images that showed only bumps on each side of Saturn.) Huygens also developed a new and more accurate clock design using a pendulum to regulate the motion of the clock. (Galileo was the first to identify the basic properties of pendulums.) Huygens helped develop the wave theory of light as well, and in 1655 on a visit to Paris, he became fascinated with the discussions among Parisian mathematicians about the mathematical theory of dice games. He did not meet Pascal, who had already abandoned math for religion; nor did he meet Fermat. He heard enough, however, to get him started with his own investigations.

One year after he had first visited Paris he completed a primer for probability. This was published in 1657. In his book, which was published in Latin with the name *De Ratiociniis in Ludo Aleae* (On reasoning in games of dice), Huygens solves a number of the same problems that had already been solved by Fermat and Pascal. He also solved some problems of his own invention. The problems are ordered and the results of previous problems are used in the solution of later ones. Again, there is no real attempt to discover the principles that underlie the problems, but Huygens's small textbook puts the new field of probability in reach of a broader audience. In contrast to the letters of Fermat and Pascal, Huygens produced a carefully written text that explains why certain statements are true and how these new ideas can be used. It is the first mathematical book written on probability, and it remained a standard introduction to the subject for about half a century.

Jacob Bernoulli

The German mathematician and philosopher Gottfried Leibniz (1646–1716) and the English mathematician and physicist Isaac Newton (1643–1727) are credited as the codiscoverers of calculus. They did not invent the entire subject on their own, however. Many of the ideas and techniques that make up calculus were already known to Fermat and others. The great French mathematician and astronomer Pierre Simon Laplace even described Fermat as the "true" inventor of the differential calculus—calculus is usually described as having a differential and an integral part—so Laplace was giving Fermat credit for discovering half the subject. There is some truth to the claim. Nevertheless, Leibniz and Newton, working independently, were the first to assemble all the disparate ideas that comprise calculus and to see them as part of a greater whole.

The impact calculus made on the mathematics of the time cannot be overstated. Many problems that were once thought difficult to solve were now perceived as easy special cases in a much broader mathematical landscape. The frontiers of mathematics were pushed far back, and for the next several generations mathematicians took full advantage of these new ideas to imagine and solve many new kinds of problems. Probability theory also benefited from the new ideas and techniques of calculus. In the theory of probability, however, Leibniz and Newton had little interest.

The Swiss mathematician Jacob Bernoulli (1654–1705) was a member of what was certainly the most mathematical family in history. Several generations of Bernoullis made important contributions to the mathematical sciences. Jacob belonged to the second generation of the mathematical Bernoulli clan, and he was one of the very first mathematicians to recognize the importance of calculus to probability as well as the importance of probability to disciplines beyond the study of games of chance. Jacob Bernoulli was educated as a minister, but ministry seems to have

76   *ARTIS CONJECTANDI*

Numerus Rerum,	Permutationum,
1 - - -	1
	2
2 - - -	2
	3
3 - - -	6
	4
4 - - -	24
	5
5 - - -	120
	6
6 - - -	720
	7
7 - - -	5040
	8
8 - -	40320
	9
9 - -	362880
	10
10 - -	3628800
	3628800
11 - -	39916800
	79833600
12 -	479001600

2. *Si rerum permutandarum nonnullæ sunt eædem :*

Quòd si literæ una pluresve recurrant sæpiùs, hoc est, si in dato rerum nume-ro aliquæ res similes sint sive eædem; ut, si datæ sint a a b c d, ubi litera a ter repetitur, numerus permutationum mul-to minor evadit: ad quem inveniendum cogitandum est, quòd, si omnes essent diversæ , putà , si loco a a a scriberetur a α a, possent hæ tres literæ etiam nul-lâ cæterarum loco motâ inter se sexies transponi, per præced. Regul. unde toti-dem diversæ nascerentur pe mutationes; at nunc cùm sunt eædem , sex istæ per-mutationes literarum a α a nullam uni-versarum dispositioni variationem indu-cunt, ac 'proinde pro unâ eâdemque ha-bendæ sunt : quod cùm de quâcunque dispositione literarum pariter sit intelli-gendum , indicium præbet , numerum permutationum rerum datarum sexies, h. e. toties minorem esse numero permu-tationum, quas subire possent si omnes essent diversæ, quoties inter se permuta-ri queunt res similes : sed si omnes 6 li-teræ diversæ existerent, permutari pos-sent juxta præced. 720. vicibus. Ergo nunc ubi tres ipsarum conveniunt, per-mutari duntaxat poterunt vicibus 120.

Iterum si datæ sint 6 literæ a a b b c, ubi præter literam a quæ ter recurrit, etiam litera b bis repetitur ; manifestum est, numerum permutatio-num adhuc bis minorem evadere, quàm in præcedenti casu fuerat, adeoque solùm ad 60 se extendere : quandoquidem binæ quælibet permu-

A page from Jacob Bernoulli's Ars Conjectandi. *In the column Bernoulli demonstrates how to compute the first few values of what is now known as the factorial function.* (Courtesy of Department of Special Collections, University of Vermont)

been his father's preference rather than his own. Instead, Bernoulli was interested in astronomy and mathematics. As is every good son, however, he was obedient to a point. He first earned a degree in theology and then left Basel, Switzerland, his hometown, and traveled around northern Europe meeting scientists and mathematicians. He exchanged ideas and learned as much as he could. At the age of 27, he returned to Basel and began his life's work as a mathematics teacher and scholar. Later, when he designed a crest (a sort of traditional seal that was popular at the time) for himself, he used the motto "Against my father's will I study the stars."

Bernoulli corresponded with Leibniz for years and developed an early interest in probability. He was especially impressed by Christian Huygens's book *De Ratiociniis in Ludo Aleae*, described earlier in this chapter. In fact, Bernoulli's major work in the field of probability, called *Ars Conjectandi*, contains a commentary on Huygens's work. (The title of Bernoulli's book translates to "the art of conjecturing," but the book is still usually referred to by its Latin name.) Bernoulli worked on *Ars Conjectandi* up until the time of his death. The book was nearly finished when he died. Jacob's nephew Nicolas finished the book after much delay, and it was published eight years after Bernoulli's death.

Many of the calculations in *Ars Conjectandi* center around games of chance. Games of chance provided a sort of vocabulary in which Bernoulli—as did Fermat, Pascal, and Huygens—expressed his ideas about randomness. But in *Ars Conjectandi* Bernoulli moves the theory of probability away from being primarily a vehicle for calculating gambling odds. He considers, for example, how probability applies to problems in criminal justice and human mortality. He did not make much progress in these areas, but it is significant that he recognized that probability theory might help us understand a variety of areas of human experience.

The most famous result obtained in *Ars Conjectandi* is a mathematical theorem called the law of large numbers, sometimes called Bernoulli's theorem. Bernoulli claims to have struggled with the ideas contained in the law of large numbers for 20 years. This mathematical discovery inspired debate among mathematicians and philosophers for more than a century after the initial publication of

Ars Conjectandi. The law of large numbers is still taught as an important part of any introductory college course on probability.

In the law of large numbers Bernoulli considered a set of random events that are *independent* of one another. In the theory of probability two events are said to be independent of one another when the outcome of one event does not influence the outcome of the other event. For example, the odds of throwing a 4 with a single die are 1/6. This is true every time one throws a die. It does not matter what one has thrown previously, because previous throws have no effect on future outcomes. Therefore, each time one throws a die the odds of throwing a 4 remain 1/6, and what can be said about a 4 can be said about any of the other numbers on the die. Mathematicians summarize this situation by saying that each throw of the die is independent of every other throw.

Next Bernoulli considered ratios *and only the ratios* that exist between the number of times a given event occurs and the total number of trials. (The reliance on ratios is important: When tossing a fair coin the difference in the total number of heads thrown versus the total number of tails will, in general, become very large provided the coin is tossed often enough. Both ratios, however, always tend toward 50 percent.) To return to dice again, Bernoulli would have considered the ratio formed, for example, by the number of times a 4 appeared divided by the number of times that the die was rolled rather than by the total number of 4s obtained:

(Number of 4s)/(Number of throws)

In *Ars Conjectandi* Bernoulli showed that *when the trials are independent*, the ratio of the number of successful outcomes to the total number of trials approaches the probability of the successful outcome. (Here the word *successful* denotes a particular outcome; it does not imply that one outcome is more desirable than another.) Or to put it another way: *If we roll the die often enough*, the frequency with which we roll the number 4 will be very close to the probability of its occurrence.

Beyond stating these observations, which may seem obvious and perhaps not even very mathematical, Bernoulli made explicit the

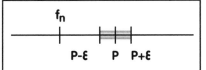

P is the probability of an event

Ɛ is our measure of "closeness."

f$_n$ is the measured frequency of an event after n trials

f$_n$ can lie within the interval for any **n,** however our confidence that fn lies with the interval about **p** increases as **n** increases

The letter p represents the probability. The Greek letter ε represents our definition of "close." Every point in the shaded interval is within ε units of p. The term f$_n$ represents the frequency of the event of interest after n trials. While f$_n$ can lie inside the interval centered at p for any value of n, our confidence that it is located within the interval increases as n, the number of trials, increases.

way in which the ratio approaches the probability of the given event. Suppose that we let the letter p represent the probability of the event in which we are interested. We can imagine a small interval surrounding p. For example, we can imagine the interval as consisting of all the numbers on the number line to the left and right of p that are within 1/1,000 of p. These numbers compose an interval with p at its center. The law of large numbers states that if the total number of trials is large enough, then the ratio formed by the number of successful events to the total number of trials will almost certainly lie inside this small interval. By *almost certainly* we mean that if we want to be 99.99 percent sure that the ratio will lie inside this interval then we need to perform only a certain number of trials. We will let the letter n stand for the number of trials we need to perform. If we throw the die n times (or more), we can be 99.99 percent sure that the ratio we obtain will lie inside the interval that we choose. Of course, there is nothing special about the number 1/1,000 or the percentage 99.99. We chose them only to be definite. We are free to substitute other numbers and other percentages. What is important is that Bernoulli made explicit an important relationship between what we observe and what we compute for a special class of random processes.

The law of large numbers made a huge impression on the mathematicians and scientists of the day. In his book Jacob Bernoulli showed that there was a robust and well-defined structure for the class of independent random processes. Although it is true that not every random process is independent, independent random processes make up an important class of random processes, and in a certain sense independent random processes are the most random of all random processes. Bernoulli succeeded in demonstrating the existence of a deep structure associated with events that until then had simply been described as unpredictable.

Bernoulli was also interested in a sort of converse to the law of large numbers. Recall that in the law we assume that we know the probability and we

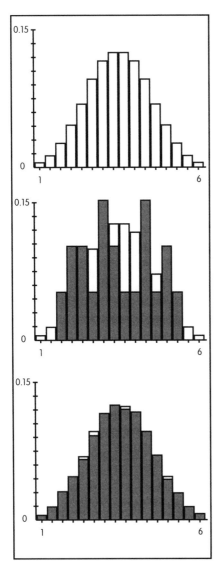

In these computer simulations three dice are rolled and the sum is divided by 3. Sixteen different outcomes are possible: 1, 4/3, 5/3, 2, . . ., 6. The top graph shows the probability of each outcome. The middle graph shows the frequency of each outcome after a trial run of 20 throws. The bottom graph shows the frequency of each outcome after a trial run of 10,000 throws. (Courtesy Professor Kyle Siegrist and The Dice Experiment, www.math.uah.edu/psol/applets/ DiceExperiment.html)

show that the measured frequency with which some event occurs tends toward the probability. Alternatively, suppose that we do not know the probability. Suppose, instead, that all we know is the relative frequency of some event after some set of trials. Bernoulli wanted to use these data to estimate the probability of the event. This is a harder problem, and Bernoulli had less success in solving it. Nevertheless, he was one of the first to recognize both halves of the same problem: (1) given the probability predict the frequency, and (2) given the frequency deduce the probability. The relationship between these two aspects of the same problem would occupy the attention of mathematicians for many years.

Bernoulli's work marks a turning point in the history of probability. His results inspired many mathematicians to attempt to apply these ideas to different problems in mathematics and science. Other mathematicians began to search for ways of generalizing Bernoulli's results. Still others debated the implications of their meaning. *Ars Conjectandi* was an important milestone in the history of probability.

Abraham de Moivre

In France, in 1667, 13 years after the birth of Jacob Bernoulli, Abraham de Moivre was born. He was a Huguenot, a French Protestant, and during this time in France the Huguenots enjoyed limited freedom under a law called the Edict of Nantes. As a teenager de Moivre studied mathematics in Paris at the Sorbonne. When de Moivre was 18, however, the edict was repealed, and de Moivre was promptly imprisoned. He remained in prison for two years. After he was released, he left for England and never returned to his native country.

Abraham de Moivre lived by his wits. His skill was his knowledge of mathematics, and he spent his adult life tutoring the rich and learning more mathematics. Largely self-taught, he first saw Newton's major work, *Principia Mathematica*, it is said, at the home of one of his students. He later purchased the book and tore out the pages, learning the entire text one page at a time as he walked about London from one tutoring job to the next.

THE

DOCTRINE

O F

CHANCES.

❀❀❀❀❀❀❀❀❀❀❀❀❀❀❀❀❀❀❀❀❀❀❀❀❀❀❀❀❀❀❀❀❀❀❀❀❀❀

The INTRODUCTION.

1. THE Probability of an Event is greater or lefs, according to the number of Chances by which it may happen, compared with the whole number of Chances by which it may either happen or fail.

2. Wherefore, if we conftitute a Fraction whereof the Numerator be the number of Chances whereby an Event may happen, and the Denominator the number of all the Chances whereby it may either happen or fail, that Fraction will be a proper defignation of the Probability

B   bability

The first page of text from de Moivre's Doctrine of Chances (Courtesy of Department of Special Collections, University of Vermont)

Over time de Moivre became friends with many of the major mathematical figures of the time, including Isaac Newton and the British mathematician and scientist Edmund Halley (1656–1742). (Edmund Halley is remembered primarily for his work in astronomy—Halley's comet bears his name—but in the second half of this book we will see that he was an important figure in the history of statistics as well.) The best-known anecdote about de Moivre involves his friend Isaac Newton. In later life, when people went to Newton for mathematical help, he would refer them to de Moivre, saying, "Go to Mr. de Moivre; he knows these things better than I do." That is quite a recommendation.

Although de Moivre made contributions in other areas, especially algebra, he is today best remembered for his work in probability theory. As Jacob Bernoulli was, Abraham de Moivre was fascinated by Christian Huygens's short work *De Ratiociniis in Ludo Aleae*. His own major work is entitled *The Doctrine of Chances, or A Method of Calculating the Probabilities of Events in Play*. Published in 1756, it is a big book and an excellent reference for understanding the state of the art in probability theory in 18th-century England. De Moivre began his book with a long introduction, which included a compliment to Huygens and a justification for *The Doctrine of Chances*. It is clear that the justification was important to him, and it is easy to see why: He was pushing back the frontiers of mathematical knowledge through the study of what many people considered a vice. Gambling, of course, is the vice, and gambling problems and their solutions are what he wanted to understand in problem after problem. He used calculus and the latest ideas from the quickly changing field of algebra. He even claimed to be developing an algebra of probability, but to a modern reader's eyes there is not much algebra in the book. Most of the problems contained in *The Doctrine of Chances* are very long because very few algebraic symbols are employed. Instead, most of the book is written out in long, carefully crafted sentences. Nevertheless, he is clearly employing the most advanced mathematics of his time in an attempt to understand an important new concept.

In a modern sense there is not much theory to de Moivre's book. Instead of theorems and proofs, de Moivre conveys his insights

DE MOIVRE ON MATH AND LUCK

Early in *The Doctrine of Chances* de Moivre dismisses the role of luck in games of chance. What he says is in stark contrast to Cardano's words on the subject. (See the sidebar Cardano on Luck and Math.) Here are de Moivre's words on the subject:

The Asserters of Luck are very sure from their own Experience, that at some times they have been very Lucky, and that at other times they have had a prodigious Run of ill Luck against them, which whilst it continued obliged them to be very cautious in engaging with the Fortunate; but how Chance should produce those extraordinary Events, is what they cannot conceive: They would be glad, for Instance, to be Satisfied, how they could lose Fifteen Games together at Piquet, if ill Luck had not strangely prevailed against them.

But if they will be pleased to consider the Rules delivered in this Book, they will see, that though the Odds against their losing so many times together be very great, viz. 32767 to 1, yet that the Possibility of it is not destroyed by the greatness of the Odds, there being One chance in 32768 that it may so happen; from whence it follows, that it was still possible to come to pass without the Intervention of what they call Ill Luck.

Besides, This Accident of losing Fifteen times together at Piquet, is no more to be imputed to ill Luck, than the Winning with one single Ticket the biggest Prize, in a Lottery of 32768 Tickets, is to be imputed to good Luck, since the Chances in both Cases are perfectly equal. But if it be said that Luck has been concerned in this latter Case, the Answer will be easy; let us suppose Luck

through a long sequence of problems. This approach to probability is reminiscent of the Mesopotamian approach to mathematics 4,000 years before de Moivre's birth. Mesopotamian scribes learned mathematics not through a study of general principles, but rather through the solution of a long sequence of problems, beginning with simple problems and continuing on to increasingly difficult ones. Similarly, *The Doctrine of Chances* begins with simple gambling problems and their solutions. As the text progresses not existing, or at least let us suppose its Influence to be suspended, yet the biggest Prize must fall into some Hand or other, not by Luck, (for by Hypothesis that has been laid aside) but from the meer necessity of its falling somewhere.

To complete the contrast between the work of Cardano, one of the best mathematicians of his day, and de Moivre, one of the best mathematicians of his day, we include a problem from *The Doctrine of Chances*. It is one of the easiest problems. As you read it, note the absence of any algebraic symbolism in the problem or in the solution. Algebra, as we know it, was still being developed and problems were often somewhat difficult to read because of the lack of algebraic symbolism. De Moivre does use some algebra, however, in expressing and solving harder problems. (By the way, the phrase *at a venture* means "to take at random.")

Suppose there is a heap of 13 Cards of one color, and another heap of 13 Cards of another color, what is the Probability that taking a Card at a venture out of each heap, I shall take the two Aces?

The Probability of taking the Ace out of the first heap is 1/13: now it being very plain that the taking or not taking the Ace out of the first heap has no influence in the taking or not taking the Ace out of the second; it follows, that supposing that Ace taken out, the Probability of taking the Ace out of the second will also be 1/13; and therefore, those two Events being independent, the Probability of their both happening will be

$$1/13 \times 1/13 = 1/169$$

more complex problems are introduced, and their solutions require ever more mathematical skill of the reader. The book ends when the problems end.

Finally, de Moivre introduces a very important and familiar idea: the bell-shaped, or normal, curve. This is a curve that has since become both a cultural icon and an important mathematical concept. The distribution of test scores, for example, has been found to be well approximated by the "bell curve." De Moivre shows that the curve has a strong connection with other, already-understood problems in probability. His discovery fits nicely into the general concept of probability as it was understood at the time, but his treatment of the curve is not a modern one. He does not use it to describe what are called *continuous distributions*, that is, sets of measurements in which the quantity being measured can vary continuously. Nevertheless, he makes important observations on the

THE BELL CURVE

The normal curve, also known as the bell curve, was first discovered in the 18th century by Abraham de Moivre. It is a useful tool for describing many random phenomena. Not every random phenomenon can be successfully described by using a normal curve, but measurements of many human activities have what is called a normal distribution, as do many sets of measurements that are generated from experiments in physics and chemistry.

Consider, for example, an Olympic javelin thrower. That person will throw the javelin often enough and consistently enough so that if we keep a record of each throw, then—over the course of many throws—the frequency with which the thrower makes various distances will be well approximated by the bell curve. (The normal distribution would not be a good approximation for the javelin-throwing efforts of most of us because we do not throw the javelin often enough. The difficulty in using the normal curve to describe our efforts is that if we practiced with the javelin every day most of us would find that our performance changed dramatically over time. By the time we accumulated a large number of measurements, the average distance and the variation about the distance would have changed substantially. This is not generally the case for Olympic athletes, who, presumably, are at the top of their game most of the time.)

To understand how the javelin thrower's efforts are approximately described by the bell curve, we need to keep in mind that the area beneath the curve is one unit. The x-axis marks the distances the javelin traveled. If we want to know the probability that the athlete will throw less than x meters we simply compute the area that is both under the curve and to the left of x. It follows, then, that the probability that the athlete will throw greater than x meters equals the area that is both beneath the curve and to the right of x.

shape of the curve and on some of its basic mathematical properties. To his credit, de Moivre clearly recognizes that he has made an important discovery, and he devotes a fair amount of space to exploring the idea and some of its simpler consequences.

Although the *Doctrine of Chances* offers no broad theoretical conclusions, it is a well-written compendium of gaming problems and the techniques and concepts required to solve them. In addition to

If we want to know the probability that the athlete will throw more than *x* meters and less than *y* meters, we compute the area beneath the curve that is to the right of *x* and to the left of *y*.

The normal curve has a number of simple geometric properties, some of which its discoverer, Abraham de Moivre, noticed immediately. The curve is symmetric about the line that is parallel to the *y*-axis and passes through the highest point on the curve. Notice that if we were to begin at the highest point on the curve and travel to the right we would reach a place where the curve descends most quickly, and then—though it continues to go down—we would descend more and more slowly. The technical name for this "breaking point" is the *inflection point*. (There is a similarly placed inflection point to the left of the highest point on the curve.) De Moivre recognized these characteristics of the curve about 250 years ago.

Since then mathematicians have learned a great deal more about the mathematical properties of this curve, and scientists have used it countless times to help them understand sets of measurements. The normal curve is the most studied and widely used curve in the field of probability.

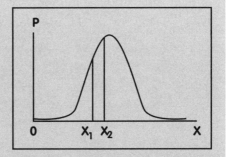

A bell curve. The area beneath the entire curve is one unit. The highest point on the curve marks the mean or average value. The probability that x lies between x_1 and x_2 equals the area beneath the curve and between the two vertical lines $x = x_1$ and $x = x_2$.

the problems associated with gambling, de Moivre studies problems of mortality from the point of view of probability. In 1756 *The Doctrine of Chances* was published along with a second text by de Moivre, *A Treatise of Annuities on Lives,* a work that depended on a paper published by Edmund Halley that analyzed birth and death rates in Breslau, a city in Central Europe. (We will have more to say about this paper in the section on statistics.)

Approximately two centuries separated Cardano's tentative musings about probability and the importance of luck and de Moivre's confident calculations and bold assertions about the nonexistence of luck. During this time Pascal, Fermat, and Bernoulli discovered new types of problems and developed important new concepts in their search for solutions. By the time *The Doctrine of Chances* was published many of the most important European mathematicians had recognized probability as a vital mathematical discipline that offered insight into a variety of problems, both theoretical and practical. This was the first new branch of mathematics to be developed since antiquity.

3

SURPRISING INSIGHTS
INTO PROBABILITY
AND ITS USES

During the 18th century, ideas about probability began to change in several fundamental ways. Previously, the theory of probability had been tied to the concepts and even the language of games of chance, but gradually mathematicians and others began to recognize the importance of probability as a tool of science. There was an urgent need for probability. The germ theory of disease was not developed until the 19th century, for example, and yet 18th-century people were dying in terrible epidemics. Choosing the best public health strategy from among several alternative strategies could not, therefore, be based solely on a detailed understanding of the biological characteristics of the disease at issue. There just were not enough facts available. Nevertheless, decisions had to be made. Mathematicians interested in public health turned to the theory of probability in an attempt to devise more effective health strategies.

As mathematicians better understood probability they discovered that it could be used to describe processes and phenomena of all branches of science. Some of their discoveries were surprising then, and they still surprise many people today. In this chapter we consider some famous examples.

Finally the definition of probability began to change as mathematicians began to think about the foundations of the subject. Previously, an imprecise idea of the meaning of probability was sufficient for the simple applications that mathematicians considered. In fact, 18th-century mathematicians were still casting about for a good definition of probability. During the latter part of the 18th century, ideas about probability began to broaden and in some ways conflict with one another. One of the earliest ideas of probability, an idea that remains both controversial and useful, was the result of the research of one Thomas Bayes.

Thomas Bayes and Inverse Probability

Thomas Bayes (1702–61) was a British minister and mathematician. He was born into a well-to-do family. As a youth he did not attend school but instead was privately tutored. Some scholars believe it likely that he was tutored by Abraham de Moivre; that possibility would help account for his skill in mathematics and his interest in probability. In any case, Thomas Bayes, as his father, Joshua Bayes, had, grew up to become a Nonconformist minister. Nonconformist ministers were religious dissidents in an age that cherished conformity. Early Nonconformists took the risk of being burned at the stake for their religious beliefs, but during the time of Thomas Bayes government oppression had diminished. In Bayes's time, refusal to conform simply meant banishment, both from public office and from the great universities of England. As a result many Nonconformist ministers were educated in Scotland or Holland, especially at the University of Leiden. Bayes was educated at the University of Edinburgh.

Bayes lived a quiet life. He worked as a minister. He corresponded with mathematicians. He was eventually elected to the Royal Society, in which he had contact with other mathematicians on a regular basis. He was a modest man who was described by his peers as a fine mathematician, but today not much is known of his life, personal or professional.

Bayes published his ideas just twice. In 1731 he published *Divine Benevolence: or, An Attempt to prove that the Principal End of the Divine Providence and Government is the Happiness of his Creatures.* In 1736 he published *An Introduction to the Doctrine of Fluxions, and a Defence of the Mathematicians against the objections of the Author of the Analyst.* Both works were published anonymously, and neither was concerned with probability theory. The first work is a religious tract that drew a lot of attention when it first appeared. The second is a defense of the fundamental ideas underlying Isaac Newton's calculus. Bayes felt compelled to write the second work because the logical foundations of the calculus had been attacked by Bishop George Berkeley in a famous work called *The Analyst.*

It is clear from the article that Bishop Berkeley felt that the scientific breakthroughs of his time were a threat to religion. Although he claims in his article that he will investigate the foundations of the subject with "impartiality," the tone of the article is hostile to the new mathematical ideas of the age. He meant his article to be controversial, and it was. Berkeley was an excellent writer, and he understood just enough about calculus to recognize which operations are fundamental to the subject. The mathematicians of Berkeley's day had found a way to employ these operations successfully to solve important problems, but they were, for the most part, still a little unclear about the mathematical basis for why the operations worked. Berkeley recognized weak logic when he saw it, and it caused him to question the validity of the entire subject. In response to *The Analyst,* Bayes attempted to express the mathematical ideas on which the calculus is founded in a more rigorous way. His goal was to prove that Berkeley's criticisms were unfounded. *An Introduction to the Doctrine of Fluxions,* as his religious writings were, was well received at the time, but neither work draws much attention today. Today, Bayes is remembered principally for a work that he never published.

When Bayes died, his family asked another minister, Richard Price (1723–91), to examine Bayes's mathematical papers. There were not many papers to examine, but there was one article about probability. That article is now known by the title "An Essay towards Solving a Problem in the Doctrine of Chances." Price recognized the importance of the work and had it published, but despite Price's best efforts Bayes's ideas attracted little initial attention. Over the succeeding centuries, however, the ideas that Bayes expressed in his manuscript have slowly attracted ever-increasing amounts of attention and controversy from mathematicians interested in probability.

It is in Bayes's paper that we find the first statement of what is now called Bayes's theorem. To understand the idea behind Bayes's theorem, imagine that we are faced with a collection of competing hypotheses. Each hypothesis purports to explain the same phenomena, but only one of the hypotheses can be the correct one. (This type of situation is common in both the physical and the social sciences.) We have no way of separating the correct hypothesis from the incorrect ones. What we do have are data. Bayes's theorem allows us to use the data *and some additional assumptions* to compute probabilities for each of the hypotheses. Just knowing the probabilities does not enable us to isolate the correct hypothesis from the incorrect ones, but it does enable us to identify which hypothesis or hypotheses are *most likely to be true.*

To make the idea more concrete, suppose that we know that there are three balls in a container. Suppose that we know that one of three hypotheses holds: (1) There are three white balls in the container, (2) there are two white balls and one black ball, and (3) there are one white ball and two black balls. Now suppose that we reach into the container and draw out a white ball. We note the color and then replace the ball. We shake the container and again we reach inside and draw out a white ball. Now we repeat the procedure a third time and again draw a white ball. Given that we have just drawn three white balls (and no black ones), and that initially we had no reason to prefer one hypothesis to the other, Bayes's theorem enables us to calculate the probability of the truth of each of the three hypotheses. Although Bayes's theorem lets us assign a probability to the truth of each of the three hypotheses, not everyone agrees that truth is a random quantity whose probability can be computed. Nevertheless, this type of probabilistic reasoning was Bayes's great insight.

Bayes's theorem is important because it allows us "to turn a probability problem around." To illustrate what this means, we introduce a little notation. Let the letter E represent some event. The probability that the event E occurs is usually written $P(E)$, and this notation is read as "the probability of event E," or "the probability of E" for short. Sometimes, however, we can make use of additional information when calculating the probability of E. Suppose, for instance, that we know another event A has occurred. We can use our knowledge of A to recompute the probability that E has occurred. This is called the conditional probability of E given A, and it is written $P(E|A)$. What Bayes's theorem gives us is a method for computing $P(A|E)$ provided we know $P(E|A)$ and some additional information. This is why we say that the probability has been "turned around": Given $P(E|A)$ we find $P(A|E)$. To be sure, we need to know considerably more than $P(E|A)$ to compute $P(A|E)$, but if we know enough then we can use Bayes's theorem to find $P(A|E)$, and this can be a very useful thing to know.

To give the matter some urgency, imagine that E represents some disease and A represents a symptom associated with the disease. We can often use a medical textbook or other source to find the probability that the patient will exhibit symptom A given that the patient has disease E. In other words, we can just look up $P(A|E)$. This is easy, but not especially helpful from a diagnostic point of view, since one symptom can be associated with several different diseases. Usually, the diagnostician is faced with the problem of determining the disease given the symptoms rather than the other way around. So what is really wanted is the probability that the patient has disease E given symptom A—that is, $P(E|A)$—and this is precisely what Bayes's theorem enables us to compute.

The algebra required to prove Bayes's theorem is neither difficult nor extensive. It is covered in almost every introductory course in probability. Perhaps it was because the math was so easy that Bayes's theorem initially escaped serious scrutiny. When a student first encounters Bayes's theorem it seems almost obvious, especially when expressed in modern notation. As time went on, however, Bayes's theorem attracted criticism from mathematicians because, as previously mentioned, Bayes's theorem requires additional assumptions. (In the ball problem, for example, we assumed that *initially* all three hypotheses were equally likely.) These extra assumptions generally involve some judgment on the part of the researcher. The researcher must make decisions about the value of certain critical terms in the necessary equations. These decisions are of a subjective nature. They are not subject to proof, and different researchers may make different decisions. (Once these decisions are made, however, the rest of the work is determined by the mathematics involved.)

Subjective judgments can be tricky because they can have the effect of introducing the researcher's own bias into the situation, and in some cases different subjective judgments can lead to very different conclusions. Some scientists object to this approach—now known as Bayesian analysis—although there are other scientists who assert that making use of the expertise of the researcher is not necessarily bad, and in any case it simply cannot be prevented. These disagreements matter because they are disagreements about how and when probability can be reliably used.

The way we understand the theory of probability helps to determine what types of problems we can solve and how we solve them. It can also affect the types of results we obtain. The ongoing disputes about the reasonableness of the Reverend Bayes's ideas are an important example of mathematicians' striving to understand the logical underpinnings of the theory of probability. Their discussions and debates about the philosophy of probability continue to reverberate throughout mathematics. Today, those scientists and statisticians who find the ideas first introduced by Thomas Bayes reasonable describe themselves as Bayesians. Those who disagree are often described as frequentists, because they prefer to work with probabilities that are determined by frequencies. The discussion between the two groups continues.

Buffon and the Needle Problem

The French naturalist and mathematician Georges-Louis Leclerc de Buffon, also known as comte de Buffon (1707–88), was the first to connect probability to a problem in geometry. As many of his contemporaries did, Buffon received a very broad education. He seems to have developed an early interest in math but originally studied law, apparently at his father's suggestion. He soon expanded his horizons through the study of botany, mathematics, and medicine but left college—or rather was forced to leave college— because of a duel. After leaving college in Angers, France, Buffon traveled throughout Europe. He lived for a time in Italy and also in England. In England he was elected to the Royal Society, but when his mother died he returned to France and settled down on his family's estate.

Buffon was intellectually ambitious. He was interested in physics, mathematics, forestry, geology, zoology—almost every branch of science—and he sought to learn as much as he could about each one. His major work was an attempt to write a series of books that would describe all of nature. It is called *Histoire naturelle générale et particulière*. In this work there is plenty of evidence of Buffon's independent thinking. He believes, for example, that the true age of Earth is 75,000 years—among his contemporaries, Earth's age was generally believed to be about 6,000 years—and he accompanied his estimate with an account of geological history. Buffon also wrote volumes about animals of all sorts, including species that had become extinct, another unusual idea for the time. Buffon developed a theory of "organic molecules." These are not molecules in the sense that we understand the term; Buffon asserted the existence of small building-block-type objects that assembled themselves into a living organism guided by some interior plan, so there is some overlap with contemporary ideas about proteins and cells. Buffon had originally planned a 50-volume set, but though he

Georges-Louis Leclerc de Buffon. His discovery that the number π can be represented as the limit of a random process drew a great deal of attention. (Topham / The Image Works)

worked furiously on the project for years, he completed only 36 volumes before his health deteriorated. His writings were both controversial and influential.

Buffon's efforts were widely recognized and respected in his own time. It is said that an English privateer captured a ship that contained numerous specimens that had been gathered from around the world and addressed to Buffon. (Privateers were privately owned vessels commissioned by one country to harass the shipping of another.) The privateers recognized Buffon's name. They probably had little appreciation for the contents of the boxes, but because of Buffon's stature they forwarded the boxes on to Paris. When Buffon died, 20,000 people turned out for his funeral.

In mathematics Buffon was interested in probability, calculus, and geometry. He wrote about calculus and its relationship to probability. He also translated a work by Newton on the calculus, but he is best known for a remarkable discovery called Buffon's needle problem. Imagine a smooth flat floor with parallel lines drawn across it. The lines are all one unit apart, and one unit can represent a foot, a meter, an inch, or a centimeter. Imagine that we have a thin rod, or needle. We will let the letter r represent the length of the needle, where we assume that r is less than one unit. (The length, r, of the "needle" must be less than the distance between two lines so that it cannot cross two lines at the same time.)

Now toss the needle at random on the floor and keep track of whether or not the needle comes to rest across a line or not. (Buffon accomplished this by tossing a rod over his shoulder.) If we let the letter h represent the number of times that the needle crosses a line and the letter n represent the total number of tosses, then by Bernoulli's theorem, as we continue to throw the needle, the ratio h/n will tend toward the probability that the needle crosses a line. What Buffon showed is that the more we throw the needle the closer the ratio h/n gets to the number $2r/\pi$. We conclude that the probability that the needle crosses the line is $2r/\pi$. Furthermore, using our ratio h/n of the number of "hits" to the number of throws, we obtain an equation that allows us to solve for π: $\pi = 2rn/h$.

This was a famous and surprising result, because it allowed one to represent a well-known and decidedly nonrandom quantity, the number π, as the limit of a very large number of random throws. Buffon's discovery pointed to still more new ideas and applications of probability. Ever since Buffon published his discovery, people have thrown needles onto lined paper hundreds or even thousands of times and kept track of the ratio h/n to observe Buffon's random process in action.

Daniel Bernoulli and Smallpox

Applications of the mathematical theory of probability to problems in science and technology can be very controversial. Although the theory of probability as a mathematical discipline is as rigorous as that of any other branch of mathematics, this rigor is no guarantee that the results obtained from the theory will be "reasonable." The application of any mathematical theory to a real-world problem rests on certain additional assumptions about

Part of a 1553 text on smallpox (Library of Congress, Prints and Photographs Division)

the relationship between the theory and the application. The mathematically derived conclusions might be logically rigorous consequences of the assumptions, but that is no guarantee that the conclusions themselves will coincide with reality.

The sometimes-tenuous nature of the connections between the mathematical theory of probability and scientific and technological applications accounts for the frequent disputes about the reasonableness of deductions derived with the help of the theory. Some of the results are of a technical nature, but others center on deeper questions about philosophic notions of chance and probability. Historically, one of the first such disputes arose when probability theory was first used to help formulate a government health policy. The issue under discussion was the prevention of smallpox. The discussion, which took place centuries ago, still sounds remarkably modern. Today, the same sorts of issues are sources of concern again. As we will later see, the discussion that began in the 18th century never really ended. It continues to this day.

Smallpox is at least as old as civilization. The ancient Egyptians suffered from smallpox, and so did the Hittites, Greeks, Romans, and Ottomans. Nor was the disease localized to northern Africa and the Mideast. Chinese records from 3,000 years ago describe the disease, and so do ancient Sanskrit texts of India. Slowly, out of these centuries of pain and loss, knowledge about the disease accumulated. The Greeks knew that if one survived smallpox, one did not become infected again. This is called acquired immunity. The Islamic doctor ar-Razi, who lived about 11 centuries ago, wrote the historically important *Treatise on the Small Pox and Measles*. He describes the disease and indicates (correctly) that it is transmitted from person to person. By the time of the Swiss mathematician and scientist Daniel Bernoulli (1700–82), scientists and laypeople alike had discovered something else important about the disease: Resistance to smallpox can be conferred through a process called variolation. To understand the problem that Bernoulli tried to solve it helps to know a little more about smallpox and variolation.

Smallpox is caused by a virus. It is often described as a disease that was fatal to about a third of those who became infected, but there were different strains of the disease. Some strains killed only a small percentage of those who became infected; other strains killed well over half of those who became ill. Those who survive smallpox are sick for about a month. There is an incubation period of about seven to 17 days, during which the infected person feels fine and is not contagious. The first symptoms are a headache, a severe backache, and generalized flulike symptoms. Next a rash, consisting of small red spots, appears on the tongue and mouth. When these sores break, the person is highly contagious. The rash spreads to the face, arms, legs, and trunk of the body. By the fifth day the bumps become raised and very hard. Fever increases. Scabs begin to form over the bumps. Sometime between the 11th and 14th days the fever begins to drop, and sometime around the third week the scabs begin to fall off. Around the 27th day after the first symptoms appear the scabs have all fallen off and the person is no longer contagious. Numerous pitted scars mark the skin of a person who has recovered from smallpox. The scars remain for life.

Before the discovery of a smallpox vaccine in the last years of the 18th century by the British doctor Edward Jenner, there were only two strategies for dealing with smallpox. One strategy was to do nothing and hope to escape infection. This strategy carried a significant risk because smallpox was widespread in the 18th century and was a major cause of mortality. Moreover, there was no successful treatment for someone who had contracted the disease. The other strategy for coping with smallpox was a technique called variolation. This was a primitive method of using live smallpox virus to confer immunity on an otherwise healthy person. Various methods of inoculation were used, but the idea is simple enough: Transfer a milder,

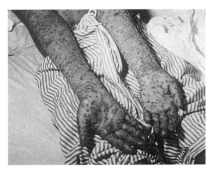

The hands and arms of a person suffering from smallpox (Courtesy of Dr. William Foege/U.S. Centers for Disease Control)

weakened form of the disease from someone who is already infected to an otherwise healthy person. The healthy person will generally become sick, but not sick enough to die. When that person recovers he or she will have acquired immunity against all future infections. In particular, the more virulent strains of the disease will pose no risk. This is how variolation works in theory. In practice, variolation has risks as well as benefits. The most obvious risk was that some of those who were variolated died of the procedure. The problem, then, was to determine whether variolation, on balance, was a better strategy than inaction in the hope of escaping infection. The answer, as it turned out, was by no means obvious.

Enter Daniel Bernoulli. He was the son of the prominent scientist and mathematician Johann Bernoulli and nephew of Jacob Bernoulli, author of the law of large numbers. A prominent mathematician in his own right, Daniel attended universities in Heidelberg and Strasbourg, Germany, and Basel, Switzerland. He studied philosophy, logic, and medicine, and he received an M.D. degree. Almost immediately after graduation, however, he began to contribute to the development of mathematics and physics. He soon moved to Saint Petersburg, Russia, where he lived for a number of years and became a member of the Academy of Sciences. Daniel Bernoulli eventually returned to Basel, where he found a position teaching anatomy and botany.

Bernoulli decided to use probability theory to study the effect of variolation on mortality, but to do so he had to phrase the problem in a way that made it susceptible to mathematical analysis. Moreover, the problem had to be more than mathematical; it had to be phrased in a way that would make his results, whatever they turned out to be, relevant to the formulation of public health policy. Suppose, he said, that a large group of infants were variolated. Those babies who survived the procedure could live their life free of the threat of smallpox. Some of the babies, however, would certainly die within a month of being variolated as a result of the procedure itself. On the other hand, if the infants were not variolated, many of them—but probably not all of them—would eventually contract the smallpox, and some of those could be expected to die of the disease. There were substantial risks associated with either strategy. Which strategy, variolation or no variolation, was more likely to benefit the public health?

In 1760 Bernoulli read the paper, "An attempt at a new analysis of the mortality caused by smallpox and of the advantages of inoculation to prevent it," to the Paris Academy of Sciences. In this paper Bernoulli summarized what evidence was available about the probability of dying of smallpox. He presented his mathematical model and his results. What he discovered is that life expectancy would increase by almost 10 percent among the variolated.

Bernoulli decided that variolation was a valuable tool for protecting the public health. He recommended it, and he was supported in this belief by many scholars and philosophers around Europe. Others disagreed. Some disagreed with his reasoning; others simply disagreed with his conclusions. The French mathematician and scientist Jean le Rond d'Alembert scrutinized Bernoulli's paper and, although he concluded that Bernoulli's recommendation for variolation was a good one, did not entirely agree with Bernoulli's analysis. D'Alembert wrote a well-known critique of Bernoulli's well-known paper. D'Alembert's response to Bernoulli's ideas illustrates the difficulty of interpreting real-world problems in the language of probability theory.

Jean le Rond d'Alembert and the Evaluation of Risk

D'Alembert was the biological son of socially prominent parents, but they did not raise him. As an infant he was placed with a local couple of modest means, and it was they who raised him. D'Alembert's biological father contributed to his son's well-being by ensuring that d'Alembert received an excellent education, but d'Alembert's loyalty was always to his adoptive parents. After d'Alembert achieved some prominence, his biological mother tried to establish contact with him, but d'Alembert had little interest in meeting her. He continued to live with his stepparents until he was middle-aged.

As with most of the mathematicians discussed in this chapter, d'Alembert's education was very broad. As a young man he learned about medicine, science, law, and mathematics. During his life he was best known for his collaboration with the French writer and philosopher Denis Diderot. Together, they produced a 28-volume encyclopedia that was one of the great intellectual works of the Enlightenment. One of d'Alembert's contributions to the project was to write most of the entries on science and mathematics.

As we have indicated previously, d'Alembert's disagreement with Bernoulli focused more on Bernoulli's interpretation of his mathematics than on the mathematics itself. His critique is a nice example of the difficulties that arise in using probabilistic reasoning. D'Alembert disagreed with Bernoulli's reasoning that an increase in the average life span of the population justified the variolation of infants. The reason is that the risk of variolation to the infant is immediate, whereas lack of variolation usually poses no immediate danger. There must be more of a balance, he argues, between the immediate loss of one's life and the possible extension of that same life. The problem, as d'Alembert saw it, is that variolation adds years to the wrong end of one's life. Essentially, he argues that it would be better to live part of one's normal life expectancy and then die of smallpox than to risk one's entire life at the outset.

D'Alembert also considered a number of related scenarios. Each scenario illustrated the difficulty of balancing the immediate risk of variolation with the longer-term risk of smallpox. He points out, for example, that Bernoulli's calculations also show that a 30-year-old man who is not variolated can expect to live (on average) to the age of 54 years and four months. A 30-year-old man who is successfully variolated can expect to live to the age of 57 years. This, argues d'Alembert, is, again, the wrong comparison to make. The risk of dying of variolation was estimated at 1/200, and it is the 1/200 chance of almost immediate death that should be of more concern to the 30-year-old than the possibility of adding a few years to the end of what was, for the time, a long life. D'Alembert questioned whether it was wise for a 30-year-old man to risk everything to extend his life to a time when, in d'Alembert's opinion, he would be least able to enjoy himself. To illustrate his point, d'Alembert asks the reader to imagine a gambler who is faced with a wager that involves a 1/200 probability of losing everything against a modest increase in total wealth. Worse, the payoff for the wager will only occur many years later, when, presumably, the winner will be less able to use the new wealth effectively. When expressed in these terms Bernoulli's reasoning seems less persuasive.

D'Alembert goes further. He offers a second, more exaggerated example. Imagine, he says, that

- the *only* cause of death is smallpox, and that variolation as an infant ensures that each person surviving the treatment lives healthily to the age of 100.

- Further, suppose that variolation carries a 1 in 5 risk of death.

- Finally, suppose that among those who forgo variolation, the average life expectancy is 50 years.

It is not difficult to show that among the variolated, the average life expectancy is 80 years. The reason is that we have to take into account that 1/5 of the target group who received treatment died in infancy. The conclusion is that the average life expectancy of the variolated exceeds that of the unvariolated by 30 years. Was the 1 in 5 chance of dying in infancy worth the additional gain? Again, is it wise to risk everything at the outset for a gain that will be realized only far in the future? In this hypothetical situation is variolation a good bet? D'Alembert responds by saying that what is good for the state—a population of healthy, long-lived individuals—is not always what is best for the individual.

D'Alembert's is an important insight, although one should keep in mind that d'Alembert eventually concluded on other grounds that variolation is still the right strategy. In particular, he says that if variolation is performed skillfully then the risk of death from the procedure can be made as small as 1 in 3,000. Since this was the average death rate for smallpox in Paris at the time, he concluded that variolation is, under these circumstances, a good bet. The risks associated with both strategies are, he says, equal, but with variolation one achieves complete protection from smallpox for the rest of one's life. D'Alembert's argument would be more convincing if he had done more research and less speculation about the actual risks involved. In any case, his article reminds us that differentiating right from wrong and true from false by using probabilistic reasoning can be both subtle and difficult.

It is interesting that both Bernoulli and d'Alembert missed what was probably the main danger associated with variolation. When a person is variolated, that person becomes sick with smallpox and can infect others. Though the variolated person may not be so sick that he or she dies, the effect on those who subsequently become infected as a result of the initial variolation is much harder to predict. They may suffer and perhaps die of a more malignant form of the disease. So in practice, though variolation benefited the individual, it might have posed too high a risk for society at large since each variolation was another possible source for a new epidemic.

Leonhard Euler and Lotteries

There were many good mathematicians during the 18th century. The discoveries of Newton and Leibniz, Fermat and Descartes, Pascal and Galileo, among others, had opened up a new mathematical landscape, and they had provided many of the conceptual tools required to explore it. Many individuals took advantage of these opportunities and made creative and useful discoveries in one or more branches of mathematics. Some of their stories are recounted in this series, but in the 18th century one individual stood out from all others. He was the Swiss mathematician and scientist Leonhard Euler (1707–83). Some histories of mathematics even call the time when Euler was active the Age of Euler.

Almost every branch of mathematics that existed in the 18th century includes a set of theorems attributed to Euler. He was unique. Many mathematicians make contributions to their chosen field when they are young and the subject and its challenges are new to them but later lose interest or enthusiasm for their chosen field. By contrast, Euler lived a long life, and his output, as measured by the number of publications that he wrote, continued to increase right to the end. In the years from 1733 to 1743, for example, he published 49 papers. During the last decade of his life, beginning in 1773, he published 355. It is worth noting that he was blind the last 17 years of his life and even that had no apparent effect on his ever-increasing output.

Euler's father, Paul Euler, was a minister with a mathematical background. His mother, Margaret Brucker, was a member of a family of scholars. Paul studied mathematics at the University of Basel. He anticipated that his son, Leonhard, would also become a minister, but this idea did not prevent the father from tutoring his

Leonhard Euler, one of the most productive mathematicians of all time (Library of Congress, Prints and Photographs Division)

young son in mathematics. These classes were enough to get the son started on a lifetime of mathematical exploration. By the time he was 13 years of age Leonhard Euler was a student at the University of Basel, and by the time he was 16 he had earned a master's degree. He studied languages, theology, philosophy, and mathematics at Basel. Later, he briefly studied medicine as well.

Euler spent most of his adult life in two cities. He lived in Saint Petersburg, Russia, where he was a member of the Academy of Sciences from 1726 until 1741. For a period of time both Euler and his friends, Daniel and Nicolaus Bernoulli, worked at the academy together. In 1741 Euler left Saint Petersburg for Berlin to work at the Berlin Academy of Sciences under the patronage of Frederick the Great. Euler was not happy in Berlin, and in 1766 he returned to Russia, where he lived the rest of his life.

Euler's contributions to probability involved the study of games of chance and research on certain specialized functions that would later play an important role in the mathematical expression of probability. With respect to games of chance, for example, Euler considered the following problem: There are two players. Each player has a deck of cards. The cards are turned up in pairs, one from each deck. If the cards are always different, then one player wins the game and the bet. If it should happen that one pair of identical cards are turned face up simultaneously, then the second player wins. Euler computed the odds of winning for each player. This kind of problem was similar in spirit to those already considered by de Moivre. Euler's best-known work on probability involved the analysis of various state lottery schemes.

While in Berlin, Euler wrote several articles on lotteries apparently at the behest of Frederick the Great. It was a common practice then as now for governments to raise money by sponsoring lotteries. One state under Frederick's control, for example, sponsored a lottery to raise money to pay off its war debts. The goal of all these lotteries was, of course, to turn a profit for the lottery's sponsor rather than the players. Euler investigated the odds of winning various types of lotteries as well as the risk that the state incurred in offering large prizes. He wrote at least two reports to Frederick on the risks associated with various schemes.

Part of the difficulty in this type of work is that these kinds of problems can be computationally intensive. That was certainly the case for a few of the problems that Euler undertook to solve. To make his work easier Euler invented the symbol $\left[\dfrac{p}{q}\right]$ to represent the expression $\dfrac{p(p-1)(p-2) \ \cdot \ \cdot \ \cdot \ (p-q+1)}{q(q-1)(q-2) \ldots 1}$, an expression that commonly arises in problems involving probability. It represents the number of ways that distinct subsets with q elements can be chosen from a set of p objects. Although the expression is now usually written as $\left(\dfrac{p}{q}\right)$ the basic notation originates with Euler.

Euler also was one of the first to make progress in the study of the so-called beta function and in hypergeometrical series. These functions play an important role in the theory of probability. The mathematical properties of these functions are not easy to identify. They are generally expressed in terms of fairly complicated formulas, and that, in part, is what makes them difficult to use. All contemporary mathematicians interested in probability acquire some skill in manipulating these functions, but Euler was one of the first to make headway in understanding their basic mathematical properties. He did not study these functions because of their value to the theory of probability, but his discoveries have found a lasting place in this branch of mathematics.

Euler's work in the theory of probability extended our understanding about games of chance, but he did not branch out into new applications of the theory. Eighteenth-century probability theory was marked by many divergent lines of thought. There was still a lot of work done on games of chance, but the new ideas were being extended to other areas of science as well. Bernoulli's work on smallpox is the most prominent example. Mathematicians and scientists were inspired by the tremendous advances in the physical sciences, and many of them tried to apply quantitative methods, and especially probabilistic methods, to problems in the social sciences. Even theology was not exempt from attempts to prove various ideas through the use of clever probabilistic techniques. The field of probability had fragmented. Many new ideas were developed during this time, but there was no unifying concept. There was no broad treatment of probability that joined all these ideas in a single conceptual framework. D'Alembert, in particular, generated a lot of heat criticizing the work of others, but criticizing others was easy because there was a general lack of insight into the underpinnings of the subject. It would be many years before the first axiomatic treatment of probability was completed.

4

RANDOMNESS IN A
DETERMINISTIC UNIVERSE

Probability theory in the 18th century consisted of scattered results, ideas, and techniques. All of these concepts came together early in the 19th century in the mind of the French mathematician and astronomer Pierre-Simon Laplace (1749–1827). Laplace's ideas about probability influenced mathematicians throughout the 19th century, and his major work in probability, *Théorie analytique des probabilités* (Analytic theory of probability), was a major source of inspiration for generations of mathematicians.

Not much is known about Laplace's early life. He was born in Normandy. Some scholars describe his parents as well off, and others have described them as peasants. Laplace himself did not talk much about his background. What is known is that at the age of 16 he entered the University of Caen to study mathematics. After a few years in university he made his way to Paris. He had acquired some letters of recommendation—it is not known who wrote the letters—and his goal was to use these letters to introduce himself to d'Alembert, perhaps the best-known Paris-based mathematician of the time.

Laplace could not have known much about d'Alembert, but he soon discovered that d'Alembert placed little value on letters of recommendation. D'Alembert refused to meet with Laplace. Fortunately, Laplace was not easily discouraged. He took a new tack. He wrote an exposition of the principles of mechanics, that branch of physics that deals with motions and forces, and he sent it to d'Alembert. It must have been an effective letter. D'Alembert quickly helped Laplace find a job teaching at the École Militaire in Paris. These were turbulent times in France. There was continuing turmoil brought about by the French Revolution and later by the military adventures of Napoléon Bonaparte (1769–1821). Laplace's friend Antoine-Laurent Lavoisier, the scientist who first formulated the principle of conservation of mass, was executed during this time. Many scientists and mathematicians found it difficult and dangerous to continue their research. Laplace, by contrast, always seemed to find it relatively easy to work. He was, for example, a friend

During the 19th century, astronomy proved to be an important source of problems for those interested in probability theory. (Courtesy of the National Aeronautics and Space Administration)

of Napoléon's for as long as Napoléon remained in power, but he was discreet about their friendship as soon as Napoléon was gone. Laplace's ability to adapt enabled him to study mathematics almost uninterrupted at a time when many of his colleagues became caught up in the furor.

Although our interest in Laplace is due to his contributions to the theory of probability, he is perhaps better remembered for his work in astronomy. But these two fields were not pursued independently. His work in astronomy contributed to his understanding of probability, and vice versa. To understand Laplace's accomplishments it is helpful to consider his astronomical discoveries in a historical context.

Laplace's astronomical work is an extension of the work of the British mathematician and physicist Isaac Newton (1643–1727). Newton developed a mathematical model to describe how the force of the Sun's gravity affected the motions of the planets. In particular, he was able to use his model to make predictions about how the planets would move under the gravitational attraction exerted by the Sun. When Newton's model was compared to what was already known about how planets move, it was discovered that the agreement between his model and the existing data was good. Newton's model was a good reflection of reality in the sense that he used it successfully to predict the motion of the planets.

Additional measurements later revealed small discrepancies between Newton's predictions and the orbital paths of the planets. The discrepancies, called perturbations, arose because of gravitational interactions between planets themselves. The continually shifting positions of the planets relative to each other made developing a mathematical model sophisticated enough to account for the perturbations difficult. This was Laplace's contribution. Laplace applied his considerable analytical talents to the problem of planetary motion and managed to account for all of the different forces involved as well as their effects. He showed that the solar system is stable in the sense that the cumulative long-term effects of the perturbations do not disrupt its structure. (This was a matter of debate at the time.) Laplace concluded that the paths of planets can be reliably predicted far into the future as well as described far into the past.

This idea of predictability was also central to Laplace's understanding of probability. Laplace had a firm belief in the concept of cause and effect. He had a deterministic view of nature. Of course, contemporary scientists also subscribe to these same ideas, but ideas of cause and effect no longer play as central a role today as they did in the philosophy of Laplace. Nowadays scientists readily concede that there are aspects of nature that are not only unknown but are also, in principle, *unknowable.* This is a very strong statement, and it is one with which Laplace would have certainly disagreed. It was Laplace's view that if one had the correct equations and one knew everything about the state of the universe for one instant in time, then one could compute all future and past states of the universe. He had, after all, already done this for the solar system.

If one accepts Laplace's idea that the universe is entirely deterministic, then there are no random processes. There are no chance outcomes. Probability theory, as envisioned by Laplace, reduces to a set of techniques required to account for errors in measurement. Uncertainty about the outcome of any process is, in this view of nature, solely a function of our own ignorance. The more we know, the less uncertain we are. In theory at least, we can eliminate all of our uncertainty provided we know enough.

Laplace wrote two works on probability. The first, published in 1812, is *Théorie analytique des probabilités* (Analytic theory of probability). This book was written for mathematicians. Another account of the same ideas, written for a broader audience, *Essai philosophique sur les probabilités* (A Philosophical essay on probability), was published two years later. These works discuss a theory of errors, theology, mechanics, public health, actuarial science, and more, and all from a probabilistic viewpoint.

In the *Essai* Laplace discusses how measurements can be analyzed by using probability theory to obtain the most probable "true" value. We have noted that de Moivre discovered the normal, or bell-shaped, curve that represents the distribution of many

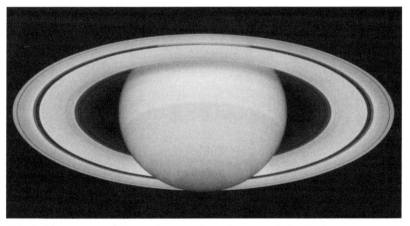

Laplace's estimate of Saturn's mass (as a fraction of the Sun's mass) was expressed in the language of probability theory in order to take into account uncertainties in the measurements on which the calculations were based. (Courtesy of the National Aeronautics and Space Administration)

random quantities. Laplace does not spend much time discussing the curve or what it represents. He evidently considered the matter fully understood. Instead, he discusses sets of measurements on the masses of Jupiter and Saturn. He analyzes these measurements from the point of view of probability theory. He computes the mean and the variance. (The *mean* is the average, or center, value of a set of measurements. The *variance* is a measure of the extent to which the measurements are dispersed about the mean.)

From his analysis he both computes the probable masses of Jupiter and Saturn and more importantly specifies limits on the accuracy of his computations. He says, for example, that the mass of Saturn is 1/3,512 that of the Sun and that the odds are 1:11,000 that his estimate is off by more than 1 percent of the computed mass of Saturn. He applies these same methods to a variety of other problems of interest to astronomers. He is interested in the problem of how inaccurate measurements can make the identification of small perturbations in planetary motion difficult to identify. He is eager to show how probabilistic methods can be used to distinguish between true perturbations and scattered, inaccurate measurements. His ability to do just that, as previously mentioned, was one of his great accomplishments in astronomy.

Laplace revisits the question of variolation as first discussed by Daniel Bernoulli and restates d'Alembert's criticisms, but by the time Laplace wrote these words the urgency of the situation had subsided. Jenner had discovered his smallpox vaccine and published his results in the last years of the 18th century. Because vaccination is much safer than variolation, the specifics of Bernoulli's analysis were of largely academic interest. Laplace goes further, however; he discusses the probable effect on the population of the elimination of a deadly disease and considers the rate at which the population will increase. This is an early attempt to come to terms with the problems of unrestricted population growth. Laplace was interested in reconciling the policy of mass vaccination and the concept of a population's increasing without limit.

Laplace also revisited and extended Buffon's needle problem. Recall that Buffon had found a way of computing the number π by randomly tossing a straight rod onto a floor marked with many equally spaced parallel lines. Buffon had discovered how to represent π as the limit of data collected during this random process. Laplace extended the problem to tossing the rod onto a floor marked by two sets of parallel lines that cross at right angles to each other. Laplace's formula is slightly more complicated, but the idea is the same. If we count how often the rod lies across one or more lines and divide that number by the total number of tosses, then we can input his information into a formula that will converge to the number π. Or, to put it in another way: The approximation to π so obtained becomes increasingly accurate as the number of tosses increases. This result has inspired many mathematicians and nonmathematicians to spend hours tossing a needle onto carefully lined paper and recording the results. It was, at the time, considered a remarkable demonstration of the power of probability—a probabilistic representation of a decidedly nonprobabilistic quantity.

There is much more to Laplace's work. For example, by Laplace's time the ideas of the Reverend Thomas Bayes had been largely forgotten. Laplace revisited Bayes's theorem, and, as he had with Buffon's needle problem, Laplace extended the work of Bayes. Recall that if we are given data and a set of explanatory hypotheses, Bayes's theorem helps us to determine which hypothesis of a competing set of hypotheses is most likely to be true. Laplace saw much more deeply into Bayes's theorem and its uses, and he explains how to use the theorem while minimizing the effect of researcher bias.

Another important contribution of Laplace is now known as the central limit theorem. The central limit theorem generalizes de Moivre's results on the normal distribution. The goal is to describe sums of random variables as the number of terms in the sum becomes large. It is a theorem that has found a wide variety of applications.

A nice example of how Laplace sometimes tried to use probability to understand science—an approach that was new at the time—is his attempt to understand atmospheric tides. Philosophers had long discussed the cause of ocean tides, although they made little headway understanding the cause of the tides until Newton proposed his law of gravity. Tides are due to the gravitational forces exerted by the Moon and Sun on Earth's oceans. The Moon has the greater effect on the tides because of its close proximity to Earth. Its gravitational attraction distorts the shape of the ocean, resulting in a regular rise and fall of the surface of the water. The cycle is repeated approximately every 12 hours. The Sun causes tides in the same general way. Its gravitational field is much stronger than the Moon's, but the Sun's greater distance makes its effect on the oceans somewhat weaker than that of the Moon.

When the Moon, Earth, and Sun are aligned, the effects of the Moon's and Sun's gravitational fields add to each other and the tides are especially high. Tides that occur under these circumstances are called spring tides, although they can occur at any time of year. When the Moon, Earth, and Sun form a right triangle, the effect of the Sun's gravitational field partially cancels that of the Moon's, and tides are generally lower. These tides are called neap tides. All of these explanations had been deduced in a rough sort of way soon after Newton had described his law of gravity.

Newton and others had speculated that if the Sun and Moon affect oceanic tides then they must affect the atmosphere as well. The effect would be subtler, but it should be measurable. Laplace set out to identify atmospheric tides, distortions in the barometric pressure caused by the gravitational fields of the Moon and Sun. Multiple measurements were made on days when the Earth, Moon, and Sun were aligned to produce spring tides, and these measurements were compared with sequences of measurements made during days when the Earth, Moon, and Sun formed a right triangle in space to produce neap tides. Since the effect of these two different geometrical arrangements could be detected in the height of the ocean, Laplace thought that it should be possible to detect their effect on the atmosphere as well.

Atmospheric pressure is affected by other factors than the relative position of the Moon and Sun, of course. Atmospheric pressure can vary quite a bit over the course of a single day, depending on the local weather. In fact, changes due to a passing high-pressure or low-pressure air mass can overwhelm any variation due to tidal effects. This is a "random" effect in the sense that the presence of a high- or low-pressure air mass cannot be predicted far ahead of time. Consequently, there were relatively large, random fluctuations in his measurements of this very subtle phenomenon. This is the reason that Laplace needed to analyze large data sets. He made several assumptions in his analysis; the one of most interest to us is that he assumed that each pressure measurement (and more than one measurement was made per day) was independent of every other measurement. This is exactly the type of assumption used by Laplace to analyze astronomical data sets, but it was a major source of error in his analysis of atmospheric tides. The atmospheric pressure in the morning is correlated with the pressure later during the day in the sense that measuring the atmospheric pressure in the morning gives us some indication of what the pressure will be later that day. The two pressures need not be the same, of course. In fact, they sometimes vary widely, but on average if we know the morning pressure we have some insight into what the barometric pressure will probably be later in the day.

In the end, Laplace's attempt to identify atmospheric tides was not successful. The effect that he was trying to identify was simply not large enough to enable him to identify it from the available data. He concluded that the differences that he did observe might have been due to chance and that to isolate the barometric effects of the Moon and Sun on the atmosphere, he would have to analyze a much larger sample.

Siméon-Denis Poisson

Probability theory arose out of the consideration of games of chance. Fair dice and well-shuffled cards have historically formed a sort of vocabulary of randomness. Sets of astronomical measurements, electoral procedures, and public health policy have all been described in terms of ideas and probabilities that are also well suited to various games of chance. There are, however, random processes that do not conform to these types of probabilities. One of the first to recognize this fact and develop another useful probability curve was the French mathematician and physicist Siméon-Denis Poisson (1781–1840).

THE POISSON DISTRIBUTION

Imagine that we want to describe the number of phone calls arriving at some destination of interest, or, perhaps, the number of automobiles passing a particular location on a busy highway. We will call the arrival of each phone call or automobile an *event*. Imagine that we observe the situation for some fixed period, which we will represent by the letter t. (The symbol t can represent a minute, an hour, a day, or a year.) The number of calls that arrive during the time interval t is random in the sense that it is

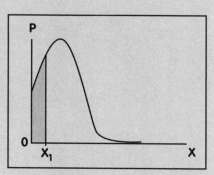

A Poisson distribution. The probability that an event will occur in the interval between 0 and x_1 equals the area beneath the curve and between the p-axis and the line $x = x_1$.

unpredictable. Now imagine that we divide the time interval t into n equal subintervals. Each small interval of time will equal t/n units. No matter how large t is, t/n will be very short, provided that we make n large enough. To use Poisson's distribution our random process must conform to three simple criteria:

Poisson was born into a family of modest means, who worked very hard to be sure that he had a good start in life. His father had a modest government position, but he supported the goals of the French Revolution, and when the revolution occurred he advanced rapidly. The family wanted Poisson to study medicine. As any good son is, Siméon-Denis Poisson was obedient to a point. He attempted to study medicine but showed little interest in the subject. Furthermore, he seems to have been remarkably uncoordinated, and that trait would have made work as a surgeon impossible. He eventually left medicine. Later, he enrolled in the École Polytechnique in Paris, where his aptitude for mathematics and science became apparent. While there, he was a student of

- For a sufficiently short period—represented by the fraction *t/n*—either one event will occur or none will occur. This condition rules out the possibility of two or more events' occurring in a single subinterval of time. This restriction is reasonable provided we choose *n* so large that the time interval *t/n* is very small, where the meaning of *large* and *small* depends on the context of the problem.

- The probability of one event's occurring in any given interval *t/n* is proportional to the length of the interval. (In other words, if we wait twice as long we will be twice as likely to observe an event.)

- Whatever happens in one subinterval (for instance, whether a phone call is received or not received) will have no influence on the occurrence of an event in any other subinterval.

If these three criteria are satisfied, then the phenomenon of interest is called a Poisson process. Once it has been established that a particular process is a Poisson process then mathematicians, engineers, and scientists can use all of the mathematics that has been developed to describe such processes. The Poisson process has become a standard tool of the mathematician interested in probability, the network design engineer, and others interested in applications of probability. It has even been used to predict the number of boulders of a given size per square kilometer on the Moon. Poisson processes are everywhere.

Laplace, who recognized in Poisson a great talent. After graduation Laplace helped him find a teaching position at École Polytechnique. Poisson was a devoted mathematician and researcher, and he is often quoted as asserting that life is good only for two things: to study mathematics and to teach it.

Poisson wrote hundreds of scientific and mathematical papers. He made important contributions to the study of electricity, magnetism, heat, mechanics, and several branches of mathematics including probability theory. His name was posthumously attached to a number of important discoveries, but he received accolades while he was alive as well. In fact, most of Poisson's contributions were recognized during his life. His peers and the broader public knew about and were supportive of his work in science and mathematics. Poisson, however, made one important discovery of interest to us that was not widely recognized during his life. This was also his major contribution to the theory of probability. It is called the Poisson distribution.

The Poisson probability distribution was first described in *Recherches sur la probabilité des jugements en matière criminelle et en matière civile* (Researches on the probability of criminal and civil verdicts). The goal of the text is to analyze the relationship between the likelihood of conviction of the accused and the likelihood of the individual's actually having committed the crime. (Estimates of this type enable one to determine approximately how many innocent people are locked away in jail. Unfortunately, they give no insight into which people are innocent.) It was during the course of his analysis that Poisson briefly described a new kind of probability curve or distribution.

Poisson's distribution enables the user to calculate the likelihood that a certain event will occur k times in a given time interval, where k represents any whole number greater than or equal to 0. This discovery passed without much notice during Poisson's time. Perhaps the reason it did not draw much attention was that he could not find an eye-catching application for his insight, but conditions have changed. Poisson processes are now widely used; Poisson distributions are, for example, employed when developing probabilistic models of telephone networks, in which they are used to predict the probability that k phone calls will arrive at a particular point on the network in a given interval of time. They are also used in the design of traffic networks in a similar sort of way. (Car arrival times are studied instead of message arrival times.) Neither of these applications could have been foreseen by Poisson or his contemporaries, of course.

5

RANDOM PROCESSES

The universe is not entirely deterministic: Not every "effect" is the result of an identifiable "cause." This idea began to find favor late in the 19th century. As scientists learned more about nature they were able to identify phenomena—for example, the motion of an individual molecule in a gas or liquid, or the turbulent flow of a fluid—for which the information necessary to identify a cause was not simply unknown but perhaps unknowable. Scientists began to look at nature in a new way. They began to develop the concept of a random, or stochastic, process. In this view of nature, scientists can specify the probability of certain outcomes of a process, but this is all they can do. For example, when studying the motion of molecules in a gas they may predict that there is a 75 percent chance that a molecule that is currently in region A will be found in region B after a given amount of time has elapsed. Or they may predict that the velocity of a turbulent fluid at a particular location at a particular time will lie within a particular range of velocities 80 percent of the time. In some instances, at least, these predictions are the best, most accurate predictions possible. For certain applications, at least, prediction in the sense that Laplace understood the term had become a relic of the past.

This kind of understanding of natural phenomena has as much in common with our understanding of games of chance as it has with the deterministic physics of Newton, Euler, and Laplace. The goal of these new scientists, then, was to state the sharpest possible probabilities for a range of outcomes, rather than to predict the unique outcome for a given cause. This was a profound shift in scientific thinking, and it began with the work of the British botanist Robert Brown (1773–1858).

Brown, like many figures in the history of mathematics, was the son of a minister. He studied medicine at the Universities of Aberdeen and Edinburgh. As a young man he led an adventurous life. He was stationed in Iceland while serving in the British army, and later he served as ship's naturalist aboard HMS *Investigator*. It was as a member of *Investigator*'s crew that he visited Australia. During this visit he collected thousands of specimens, and on his return to England he set to work classifying the collection and writing about what he found. In 1810 he published part of the results of his work as naturalist, but, because sales of the first volume were meager, he never completed the project.

Today, Brown is remembered for his observations of the motion of pollen in water made many years after his return to England. In 1828 he described his discoveries in a little pamphlet with the enormous title "A brief account of microscopical observations

Simulation of two-dimensional Brownian motion

made in the months of June, July and August, 1827, on the particles contained in the pollen of plants; and on the general existence of active molecules in organic and inorganic bodies." In this work Brown describes what he saw when he used a microscope to observe pollen particles that were about 0.0002 inch (0.0056 mm) in diameter immersed in water. He saw the particles occasionally turning on their axis and moving *randomly* about in the water. Prolonged observation indicated to him that the movements were not caused by currents or the evaporation of the water. At first, Brown referred to Buffon: He assumed that the particles moved because of the motion of the "organic molecules" whose existence had been described in Buffon's *Histoire naturelle générale et particulière*. Further research, however, changed his mind. Brown observed the same phenomenon with particles that could not be alive. He observed 100-year-old pollen. He ground up glass and granite and observed that the particles moved through the water just as the pollen had. He even observed ground-up fragments of the Sphinx. Every sufficiently small particle suspended in water behaved in essentially the same way: (1) Each particle was as likely to move in one direction as in another, (2) future motion was not influenced by past motion, and (3) the motion never stopped.

That the motions might indeed be random was not a popular hypothesis. Scientists of the time believed that these motions would eventually be explained by some yet-to-be-discovered deterministic theory much as planetary orbits had already been explained. Brown, however, continued to gather data. He was remarkably thorough. When it was suggested that the motions were due to mutual attraction between particles, he observed single grains suspended in individual droplets that were themselves suspended in oil. The oil prevented evaporation of the water, and the continued motion of isolated grains disproved the hypothesis that the motion was caused by forces between particles. Through his experiments Brown gained considerable insight into what did *not* cause the motion of these grains, but no one at the time had a convincing theory of what did cause their motion. Interest among his contemporaries, never strong to begin with, began to wane. For the next 30 or so years Brown's experiments, which described the process now known as

Brownian motion, were pushed aside. Scientists were not yet ready to consider fundamentally random events.

James Clerk Maxwell

A pioneering attempt to consider a phenomenon governed by the rules of chance occurred in 1876 in a paper published by the British physicist James Clerk Maxwell (1831–79). Maxwell was born into a middle-class family. His mother died of cancer when he was nine. He was tutored at home for a while and later attended various schools. From an early age he was something of a freethinker. He paid little attention to exams, but he published his first paper when he was only 14. That paper was on mathematics. Maxwell enjoyed mathematics from an early age but would never become a great mathematician. In fact, he published articles that contained incorrect mathematics. His future was not in mathematics; it was in science, and in science his physical insight was second to none. He is widely regarded as the most important physicist of his century.

James Clerk Maxwell, one of the most prominent scientists of the 19th century (Science Museum, London/Topham-HIP/The Image Works)

Maxwell is important to the history of probability because he discovered a new and important use for probability. This did not involve much in the way of new mathematics. Instead, Maxwell found a new application for existing mathematics: He used probability theory in the study of gases. To understand his contribution, we need to keep in mind that the atomic theory of matter was still open to debate during Maxwell's lifetime. Maxwell supposed that every gas was composed of molecules that were in constant motion. He supposed that they frequently collided with one another. The collisions, of course, changed both the direction and the speed of the molecules involved, but Maxwell went much further than this simple observation. To understand Maxwell's model, consider the following:

- Imagine large numbers of small, widely separated particles that collide elastically (that is another way of saying that when they collide, they change direction and speed but produce no heat). A good model of an elastic collision is the collision of two billiard balls.

- Suppose that these molecules are enclosed in a container in such a way that they are completely isolated from the surrounding environment (again, in much the same way that billiard balls roll on a level, smooth billiard table).

- Finally, imagine that when these particles collide with the walls of the container, these collisions, too, are elastic.

Maxwell discovered that the velocity with which each molecule is moving at some instant can vary widely from molecule to molecule, but the *probability* that a particular molecule's velocity falls within a given range at a certain instant *can be predicted*. This function, which enables one to determine the probability that the velocity of a randomly chosen molecule lies in some range, is called the velocity distribution. The idea is easier to appreciate, however, when expressed in terms of speeds: Given two speeds, s_1 and s_2, with s_1 less than s_2, the speed distribution enables one to determine the probability that the speed s of a randomly chosen molecule is greater than s_1 and less than s_2. Furthermore, in this model the physical properties of the gas, such as its pressure, can be obtained from the *average* (random) motion of the molecules of which the gas is composed. The discovery that the velocities and speeds of the individual particles followed a certain type of

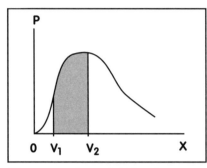

To compute the probability that a randomly chosen molecule will have a speed not less than v_1 and not greater than v_2, find the area beneath the curve and between the vertical lines $v = v_1$ and $v = v_2$.

probability distribution enabled him to describe many of the basic physical properties of gases provided that the gases under consideration are at low pressure and high temperature.

Here was a profound physical theory about the motion of individual particles that could be explained only in the language of probability. It was hopeless to describe the motions of trillions of molecules with the deterministic approach that had characterized most scientific inquiry since Newton. To use Newton's laws to describe the motion of a body we need to know the position and velocity of each body at some instant of time. Maxwell's kinetic theory of gases recognizes that the individual motions of the molecules are too complex to be described that way, but more importantly Maxwell recognized that properties of the gas are *group* properties. There are many arrangements of individual molecules in a gas that cause the same pressure and temperature. Maxwell recognized that the motions of the individual molecules are less important than the properties of the mass as a whole. Therefore, what was required to understand the gas was a probabilistic description of the motion of the molecules as a whole, not a deterministic description of each individual molecule. Furthermore, Maxwell showed that in the case of certain gases, the velocities of the molecules of the gas have a comparatively simple probabilistic description: The velocities of molecules in a gas that conform to the three "axioms" listed previously conform to something called the Maxwell–Boltzmann velocity distribution. The probabilistic description of gases that grew out of Maxwell's investigations is now known as the Maxwell–Boltzmann distribution law, after Maxwell and the

Austrian physicist Ludwig Boltzmann, the two scientists who contributed most to the development of these insights.

Brownian Motion Revisited

As scientists became accustomed to thinking of nature in the language of probability, a new, qualitative description of Brownian motion began to evolve. The general idea is that a particle that is small and light and continually bombarded by molecules traveling at varying velocities will experience forces of different magnitudes at different points along its surface. Sometimes these forces will be equally distributed about the exterior of the particle. When this happens the forces balance, or cancel, each other. When the forces do not cancel, the forces exerted on one side of the particle will exceed those exerted on the other side and the particle will "jump" from one position in the water to the next. Because the location on the surface at which the stronger forces are exerted is random, the particle may first jump in one direction one instant and then in a different direction the next, or it may jump several times in the same general direction. Understanding the cause of Brownian motion, however, was not enough to enable scientists to make quantitative predictions about the motion of the particle.

A quantitative explanation of Brownian motion was proposed independently by two scientists at essentially the same time. One was the German physicist Albert Einstein (1879–1955). He published his paper in 1905. The second person to propose the correct quantitative explanation for Brownian motion was the Polish physicist Marian Smoluchowski (1872–1917), who published his paper in 1906. Smoluchowski, who was educated at the University of Vienna, made contributions in several areas of physics. Perhaps his best-known contribution was to the understanding of Brownian motion. Unfortunately, that contribution is eclipsed by that of the much better known Albert Einstein, but they arrived at essentially the same conclusions by using different methods of reasoning at essentially the same time. We will study Smoluchowski's ideas because from our point of view the approach taken by Smoluchowski is more accessible.

Albert Einstein published an important paper on the nature of Brownian motion. (Topham/The Image Works)

Smoluchowski begins his paper on Brownian motion by reviewing previous theories about the nature of the phenomenon. Since Brown had first published his observations, a number of theories had been proposed to explain the motion of the particles. One popular theory held that small convection currents inside the fluid simply carry the particles from one location to the next. Smoluchowski cites existing evidence that disproved this possibility. In the same way, he carefully describes and dismisses all common, competing theories.

When Smoluchowski finally begins to describe his own ideas, he expresses himself in a language that would have been familiar to de Moivre. Of course, the mathematics had advanced considerably, but the concept had its roots in games of chance. Essentially, he considers a process—which for the purposes of computation is equivalent to a game—with two equally likely outcomes, favorable and unfavorable (winning/losing). The goal is to compute the odds that after running the process (playing the game) n times, where the letter n represents any positive whole number, the observer will witness m favorable outcomes, where m is any nonnegative whole number less than or equal to n.

Smoluchowski calculates an average velocity of the particle from the average velocities of the molecules that surround it. Smoluchowski's model predicts a particle that is in continual motion along a very specific random path. The path is like a chain. All steps, or links in the chain, have identical length, but the direction of each step is random. When *m* steps, where *m* is some number greater than 1, happen to line up so that motion of the particle is more or less in one direction, we have the same sort of situation that Brown had observed 79 years earlier: a particle "darting" first in one direction and then in another while the lengths of the jumps vary.

Brownian motion has since become an important part of mathematical analysis and science, but even before these ideas found practical applications, they were still recognized as important. These results helped reveal a new aspect of nature, an aspect in which randomness *could not be ignored*. This was new, because although Maxwell had developed a model for gases that relied on probability theory, a model that revealed a new way of looking at nature, there was a competing model, called the continuum model, that did not rely on any notion of randomness and was just as accurate as Maxwell's model in practical problems. From the point of view of applications, Maxwell's model was interesting, but it was not necessary, in the sense that the physical properties of gases could not be predicted without it. By contrast, the motion of a particle suspended in a fluid and battered by the surrounding molecular medium is inherently random. No nonrandom theory could account for it. Brownian motion defies the sort of cause-and-effect analysis that is characteristic of the science of the 19th century. To analyze these new classes of random phenomena new mathematical tools were needed. Brownian motion was the beginning. The need for probabilistic models in the physical sciences has continued to grow ever since.

Markov Processes

At about the same time that Smoluchowski was pondering Brownian motion the Russian mathematician Andrey Andreyevich Markov (1856–1922) had ceased studying number theory and analysis, that branch of mathematics that grew out of calculus, and begun to think about what we now call random or stochastic processes. Born in Ryazan, a city about 100 miles (160

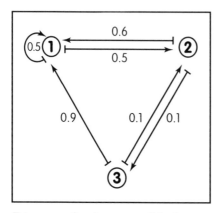

Diagram of a three-state Markov process

km) southeast of Moscow, Markov spent most of his life in Saint Petersburg. He had a comfortable childhood, but he was not a very apt student. In fact, he seems to have been a poor student in every subject except mathematics. It was by choice: Markov always did as he pleased; throughout his life he was something of a rebel.

Markov was educated at Saint Petersburg University, a home to many distinguished mathematicians then and now. Later Markov taught at Saint Petersburg. During this time the political situation in Russia was unstable and dangerous. Markov allied himself with the dissidents and against the czar, and he found a way to incorporate his mathematics into his politics. When the czarist government organized a celebration of the 300th anniversary of the Romanovs, the ruling family of Russia, Markov organized a countercelebration of the 200th anniversary of the release of Jacob Bernoulli's great work on probability theory, *Ars Conjectandi*.

In probability Markov contributed to the understanding of the central limit theorem and the law of large numbers, also called Bernoulli's theorem. His best-known contribution is his research into the study of a class of random, or stochastic, processes called Markov chains. (When a mathematical discovery is named after a mathematician, the name is usually chosen by others, often after the death of the individual responsible for the discovery. It was characteristic of Markov that he named the Markov chains after himself.) Markov chains have the following three important properties:

- The Markov chain or process is a sequence of random events.

A MARKOV CHAIN

Mathematically, a Markov chain is the simplest of all Markov-type processes. As an illustration of a Markov chain, imagine a particle moving back and forth along the real number line in discrete steps. Imagine that the particle begins its motion at the point 0 and that it can move only one unit of distance to the left or to the right at each step. In particular, after its first step it will be located at either $x = 1$ or $x = -1$. Suppose that the probability that it moves to the right at each step is p, and the probability that it moves to the left is $1 - p$, where p is any number between 0 and 1.

If we set p equal to $1/2$ then we can model this Markov process by moving back and forth along a line, one step at a time, and using a coin to determine in which direction to step: "Heads" indicates a step forward. "Tails" indicates a step back. Flip the coin. Take the step. Repeat forever. This is a simple mathematical model for Brownian motion in one dimension.

Having used the Markov chain to create a mathematical model of one-dimensional Brownian motion, we are now in a position to begin a quantitative examination of it. We might, for example, ask what is the probability that we will remain in some interval centered about the point 0 after we have taken n steps? Or, alternatively, what is the probability that after we flip the coin enough times we will move away from 0 and never return?

This simple Markov chain has been studied extensively. Although it is very simple, it has a number of more sophisticated extensions. Physically, these extensions can be used to study the phenomenon of *diffusion*, the process by which different gases or liquids intermingle as a result of the random motion of the molecules. Diffusive processes also occur in the life sciences; for example, the motion of species across the landscape is sometimes described by using diffusion equations. Mathematically, one-dimensional Markov chains have been generalized in a variety of ways. The most obvious generalization enables the particle to move in two or more dimensions. A more subtle generalization involves changing the model so that the particle moves continuously through time and space: In other words, there are no discrete steps; the particle flows randomly from one position to the next. Motions of this type are called continuous Markov processes. The study of Markov chains and continuous Markov processes continues to occupy the attention of mathematicians and scientists.

- The probability of future events is determined once we know the present state.

- The probability of future events is not influenced by the process through which the present state arose.

To predict the future position of a particle undergoing Brownian motion these are just the assumptions we would need to make. The sequence of random events is just the sequence of steps, or "links," in the Markov chain that the particle traverses as it moves along its random path. The probability that it will pass through a future location is determined by its present position and the probabilities, called transition probabilities, that govern its motion. The path that it took to arrive at its present location has no influence on its future motions.

Markov's interest in stochastic processes was very general. He did not develop his ideas in response to the problem of Brownian motion, although his ideas have been successfully applied to the study of Brownian motion. Much of Markov's motivation stemmed from his desire to make probability theory as rigorous as possible. Although he apparently enjoyed thinking about applications of his work, he made only one attempt to apply his ideas on probability: He made a probabilistic analysis of some literary works in which he modeled them as random sequences of characters. More generally, Markov worked to discover the mathematics needed to describe classes of abstract random processes. Over the last century his ideas have become a vital part of many branches of science. Today, Markov processes are used to describe stock market behavior, numerous problems in the biological and social sciences, and, of course, Brownian motion. They are also fundamental to the study of the theory of digital communication.

6

PROBABILITY AS A
MATHEMATICAL DISCIPLINE

The discoveries of Einstein, Smoluchowski, and Maxwell were just the beginning. During the first few decades of the 20th century it became apparent that there were many phenomena that could only be described using probability. It could be proved that some of these phenomena were intrinsically random. There was no choice but to use probability theory in creating a mathematical model of these processes. Laplace's philosophy, that the universe was deterministic and that the principal role of probability was to aid in the analysis of collections of measurements, had been found wanting. It is not that Laplace and others were wrong, but that their conception of nature and of the role of probability in the description of nature was too limited. Mathematicians and scientists needed a broader, more useful definition of probability.

Scientists in fields ranging from meteorology to theoretical physics had only limited success in using probability, however, because, from a mathematical perspective, the theory of probability was seriously deficient. Although many new concepts and computational techniques had been developed since the time of Jacob Bernoulli and Abraham de Moivre, there was no conceptual unity to the subject. Probability was still a haphazard collection of ideas and techniques. The time was right again to ask the question, What is the mathematical basis of probability?

It may seem that discovering a mathematical basis for the theory of probability should have been one of the first goals in the development of the subject. It was not. One reason for the delay in confronting this fundamental question was that the naïve ideas about probability that grew out of the study of games of chance had been adequate for solving many of the problems that mathematicians considered for the first few centuries after Pascal and Fermat. Another reason for the delay is that the mathematics necessary to construct a strong foundation for the theory of probability is fairly advanced. Before the 20th century the necessary mathematics did not exist. It was during the first part of the 20th century that the mathematics needed to express these fundamental ideas was first developed. The mathematicians who prepared the groundwork for a more rigorous study of probability were not especially interested in probability themselves, however; they were interested in the problem of measuring the volume occupied by arbitrary sets of points. It was during the early part of the 20th century that the French mathematicians Emile Borel (1871–1956) and Henri-Léon Lebesgue (1875–1941) revolutionized much of mathematics with their ideas about measure theory.

Andrei Nikolayevich Kolmogorov. He placed the theory of probability on a firm mathematical foundation for the first time. (Yevgeni Khaldei/CORBIS)

Measure theory is a collection of ideas and techniques that enable the user to measure the volume or area occupied by sets of points. It is a simple enough idea: We isolate the collection of points in which we are interested and then use the necessary mathematics to determine the area or volume occupied by the collection. Measure theory is closely related to integration, which is an important concept in calculus, and centuries earlier Isaac Newton and Gottfried Leibniz, the codiscoverers of calculus, had developed many important ideas and techniques with respect to integration. In the latter half of the 19th century, however, problems arose where the concepts and techniques pioneered by Newton and Leibniz proved to be inadequate. The old ideas were just too narrow to be of use in the solution of these new problems. Integration, one of the fundamental operations in all of mathematics, had to be revisited, and the concepts and techniques had to be expanded to meet the needs of the new sciences and mathematics. This was the great accomplishment of Borel and especially of Lebesgue, who found a way to extend the classical ideas of Newton and Leibniz. All of the old results were preserved and the new concepts and techniques were brought to bear on situations that previously had been unsolvable.

The fundamental work of Lebesgue and Borel was largely complete when the Russian mathematician Andrei Nikolayevich Kolmogorov (1903–87) began to think about probability. Kolmogorov was one of the major mathematicians of the 20th century, and his ideas about probability have done more to shape the subject into what it is today than any other mathematician's. He was born in Tambov, a city located about halfway between Moscow and Volgograd (formerly Stalingrad). Kolmogorov enrolled in Moscow State University when he was 17 and was soon working on problems in advanced mathematics. Eight years later he graduated from Moscow State University and joined the faculty. Shortly after he joined the faculty he began to think about probability. Kolmogorov would continue to research probability and related problems for the rest of his life. (Kolmogorov's interests were actually much broader than the field of probability. He was a prolific

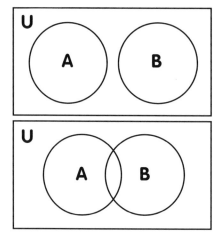

Venn diagrams. Kolmogorov discovered how to express his ideas about probability via set theory and measure theory and gave rigor to the theory of probability.

mathematician and also contributed to our understanding of the theory of complexity, information theory, turbulence, Markov processes, measure theory, geometry, topology, set theory, and other areas besides.)

Kolmogorov found a way to apply measure theory, pioneered by Borel and Lebesgue, to the study of probability theory. The idea is, in retrospect, simple enough. He imagines a large set, which we will represent with the letter *U*. The set *U* contains many subsets. On the set *U* Kolmogorov defines a measure that allows him to determine the size of various subsets of *U*. The measure must be chosen so that the size of *U* equals 1. All that is left is to reinterpret this model in the language of probability theory.

The set *U* represents all possible events or outcomes for the process of interest. (This is why it is critical that the measure of *U* equals 1: The probability that some outcome occurs—that is, that *something* happens—is always 1.) Subsets of *U* represent possible events. Because no subset of *U* can have a measure larger than *U*—a part is never larger than the whole—the probability of an event is never greater than 1. If *A* and *B* are two nonoverlapping subsets of *U*—and now *A* and *B* also represent two events—then the probability that either event *A* occurs *or* event *B* occurs is just the measure of *A* plus the measure of *B*. Geometrically, this is the size of the two sets. Alternatively, if we want to know the probability that both events *A and B* occur then we just compute the size of the intersection of the two sets. (See the accompanying illustration.)

Kolmogorov's insight allowed him to bring the field of probability into the much larger and more highly developed branch of mathematics called analysis, of which measure theory is only a part. Analysis, which arose out of calculus, is the study of functions, equations, and sets of functions. By expressing probability in the language of measure theory all of the results of analysis could now be applied to probability. The practical effect of Kolmogorov's work was widespread and immediate. Scientists and mathematicians began to employ probability theory in new ways. From a practical point of view Kolmogorov's innovation stimulated the use of probability as a tool in the study of the atom, meteorology, and the motion of fluids with internal structure, such as liquids with bubbles or liquids with solids suspended in them. (The study of such fluids has many important practical applications, ranging from coolants in nuclear plants to the motion of oil through soil.) Mathematically, Kolmogorov's innovation allowed mathematicians interested in probability to axiomatize their subject in much the same way that Euclid had attempted to axiomatize geometry more than two millennia earlier. That is, Kolmogorov was able to state the mathematical basis of the theory of probability in terms of a list of fundamental properties called axioms.

Axioms define what it is that mathematicians study. Each branch of mathematics is defined by a set of axioms. It is from the axioms that mathematicians deduce theorems, which are logical consequences of the axioms. The axioms are the final answer to the mathematical question, Why is this true? Any statement in mathematics is, in the end, true because it is a logical consequence of the axioms that define the subject. Because mathematics is a deductive subject—mathematicians draw specific logical conclusions from general principles—Kolmogorov's axiomatic approach allowed probability to be developed in a mathematically coherent way. Although others had tried, Kolmogorov was the first who successfully created an axiomatic basis for probability. Kolmogorov provided a framework that allowed those mathematicians who accepted his axioms to deduce theorems about probability rigorously. One especially important advantage of Kolmogorov's work was that it allowed probability to be applied to situations that were very abstract, situations that had previously resisted analysis.

Kolmogorov made other contributions to probability as well. In particular, he greatly expanded the results of Markov. Kolmogorov's extensions of Markov's work facilitated the study of Brownian motion and, more generally, the process of diffusion. (Diffusion occurs when molecules or particles in random motion intermix.) The mathematics of diffusion has been an important tool in the study of many problems in physics, chemistry, and certain aspects of the life sciences.

Kolmogorov also contributed to branches of knowledge in which probability theory plays an important part. One application of probability in which Kolmogorov had a particular interest was the field of information theory, the study of certain fundamental principles that govern the transmission and storage of information. The discipline emerged shortly after World War II with the work of the American engineer Claude Shannon (1916–2001). Kolmogorov developed a somewhat different approach to information theory that shared ideas with those first developed by Shannon but was more general in concept. Especially interesting and potentially useful was his idea of the information content of abstract mathematical sets. In particular, Kolmogorov found a way to compute the amount of information that could be represented by a function or group of functions whose properties are imprecisely known. Since all measurements are imprecise, this method has clear applications to the problem of interpreting data. His information theoretic ideas generated many interesting and important papers, especially by the mathematicians of the former Soviet Union.

The other contribution of Kolmogorov we mention here is a field of mathematics that is now named after him: Kolmogorov complexity. Kolmogorov complexity involves the attempt to quantify the amount of complexity in an arbitrary object. That object may be a physical object or a computer program. The first step involved in this process is to describe the object or process in binary code. Once this has been done for two objects we have two strings of digits, both of which are expressed in a common "alphabet," that is, binary code. At this point the strings are compressed, meaning that all of the information is retained but all of the redundant digits and sequences of digits are removed. At this point the strings have been made as short as possible. Now researchers are, in principle, able to compare the complexity of the two objects by using objective criteria. An objective criterion for complexity and information content is useful because random sequences of digits cannot be compressed. Consequently, the theory of complexity allows the user to distinguish between information and noise— that is, signals that carry no information—even when the message itself is opaque. If this sounds too abstract to be useful, it is not. Kolmogorov complexity has been used to gain insight into everything from the trading of stocks—in which the "signals" are fluctuations that reliably presage market changes, and the "noises" are random fluctuations that presage nothing—to the structure of language. More work remains to be done, however, because in practice, the difficulties involved in finding a reasonable measure of complexity have not been entirely resolved.

It should be pointed out that, although Kolmogorov had a profound influence on the development of the subject of probability, not every mathematician has found his formulation of the subject useful enough. We have mentioned this problem in the Introduction and we mention it again here, near the end of our history of probability, because disputes about the meaning of probability continue. Kolmogorov found a way to axiomatize probability *as he understood it.* Although other mathematicians and scientists have not always found his formulation of the subject adequate for their needs, this does not mean that Kolmogorov was wrong. He was, for Kolmogorov, right: If we accept his axioms then we must accept all of the statements about probability that are logical consequences of his axioms. Kolmogorov's highly abstract formulation, however, offers few clues to the relationship between his probability measures and the concrete, random processes that scientists seek to understand through observation and measurement. In other words, the relationship between Kolmogorov's mathematical formulation of the theory of probability and the world around us is not always clear. This is a problem of which Kolmogorov himself was aware. His formulation of the theory of probability helps us to understand how to manipulate probabilities, but it provides less insight into how we obtain probabilities, or what the probabilities so obtained can reasonably represent.

Theory and Practice

The debate about the relationship between mathematical probability and random phenomena is a lively one that continues into our own time. There is a natural tension between the mathematics of the theory of probability and real-world phenomena. The fact that the debate has lasted so long indicates that the issue is a complex one. Essentially, the question centers around the connection between what we see and what we compute.

The relationship between data and probabilities has been a matter for research and debate ever since Jacob Bernoulli demonstrated that over the long run the frequency of occurrence of a random event will approach the probability of that event. Although it was an important first step, Bernoulli's discovery was far from the last word on the subject. In order to use probability to solve problems in science, scientists needed to identify other, deeper relations between the quantities that they measure and the probabilities with which they make computations. Because these relationships are not always obvious, there are different schools of thought on the nature of probability. Identifying the "right" relationships between theory and practice is important because in more complicated research situations, the answers researchers obtain sometimes depend on their concept of probability. This variation from researcher to researcher calls into question the validity of their results. The physical properties of any phenomenon are, after all, the same no matter which method is used to reveal those properties.

One group of mathematicians interested in probability calls its members Bayesians. This school of thought goes back to the ideas of Bayes and Laplace. Bayesian probability is sometimes described as a measure of one's certainty. Briefly, Bayesians tend to ask the most natural-sounding questions about a process. They are also

Before the launch of the first space shuttle, estimates of a catastrophic failure indicated that it was very unlikely that even a single craft would be lost over the course of the program. (Courtesy of the National Aeronautics and Space Administration) free to ask questions that frequentists, those mathematicians who subscribe to a different point of view (described later), would never ask. To illustrate this point, consider the question, What is the probability that humans will land on Mars before 2050?

The question may sound reasonable, but we cannot estimate the *probability* of such a mission by examining the number of previous manned Mars landings, because this kind of mission has never occurred. In fact, this kind of mission has never even been attempted. Consequently, if we are to answer the question at all, we have

The astronaut Buzz Aldrin. How can we compute the probability of arriving safely at a place no one has ever been? (Courtesy of the National Aeronautics and Space Administration)

to talk about probabilities separately from frequencies. We are left with the problem of trying to quantify our degree of belief. Methods to facilitate this type of approach exist, but not every mathematician agrees that the methods are valid. The idea is to consider a collection of competing hypotheses. The hypotheses should be exhaustive in the sense that one and only one hypothesis can be true. The next step is to use existing data or other theoretical considerations to assign numbers to the competing hypotheses. Each number represents a *degree of plausibility*. As more information is gained, the new information can be used to "update" the initial probabilities: Knowing more enables us to decrease our level of uncertainty. This part of the calculation can be done in a rigorous way, but the initial probabilities depend on the subjective judgment of the researcher. As more information is added to the model, the probabilities change and, presumably, improve. The method yields one or more hypotheses that are the most likely or—to put it another way—the most plausible. If there is a single hypothesis that is the most plausible, then this hypothesis is accepted as the correct one—pending, of course, the introduction of additional information into the equations. There is nothing new about this understanding of probability. Mathematicians have been familiar with it in one form or another for centuries. Ideas about probability began to change, however, in the second half of the 19th century with the work of the British priest and mathematician John Venn (1834–1923).

Venn was one of the originators of what has become known as the frequentist's view of probability. He attended Gonville and Caius College, Cambridge. After graduation he was ordained a priest, but he was soon back at Cambridge, where he worked as a lecturer in moral science. Today, Venn is best remembered for the diagrams he invented to represent operations on sets—we used his diagrams to describe Kolmogorov's ideas on probability—but Venn was just as interested in the theory of probability. In 1866 Venn introduced what is now called the frequentist definition of probability. Venn's goal was to connect the ideas of probability and frequency. He defined the probability of an event as its long-term frequency. This concept of probability, although it sounds reasonable, is not quite correct. Defining probabilities in terms of frequencies turned out to be more difficult than it first appeared.

The shortcoming in Venn's conception of probability stemmed from his omission of the concept of randomness. Long-term frequencies are not enough. To see the problem, consider the sequence 5, 6, 5, 6, 5, 6, . . ., which consists solely of alternating 5s and 6s. The frequency of a 5 is 50 percent, but after we know one number, we know the order of every other number in the sequence, so, in particular, if we observe a 6, the probability that the next number is a 5 is not 50 percent; it is 100 percent.

This shortcoming in Venn's definition was eventually corrected by the mathematician Richard von Mises (1883–1953). He was born in what is now Lvov in Ukraine, which was then part of the Austro-Hungarian Empire. Von Mises was educated in Vienna before World War I, and during the First World War he served as a pilot. After the war he moved to Germany, where he taught mathematics. In the 1930s von Mises fled to Turkey to escape Nazi persecution. In 1938, when the Turkish leader Kemal Atatürk died, von Mises moved to the United States, where he lived until his death.

Von Mises recognized that long-term frequencies alone were not enough to establish a concept of probability. He had the idea of adding the requirement that a sequence must also be random in the sense that we should not be able to use our knowledge of past events to eliminate all uncertainty about future outcomes. For example, in a numerical series consisting of 50 percent 5s and 50 percent 6s, it should not be possible to predict upcoming digits with complete accuracy even if we know everything about the preceding digits. This emphasis on frequencies of randomly occurring digits "over the long run" provided an alternative to the Bayesian approach. An alternative was needed because many mathematicians of the time objected to Bayesian probability.

One major philosophical objection to the Bayesian view was that "degree of plausibility" of a hypothesis seemed to be a fairly thin reed on which to construct a rigorous mathematical discipline. Other objections were technical. Technically, the Bayesian emphasis on the calculation of probabilities of competing hypotheses depended on certain subjective judgments by the researcher. These subjective decisions seemed (to the frequentists) to leave little room for the development of an objective science. They argued that Bayesian probability ought to be replaced with a more "objective" approach.

This so-called objective approach, the frequentist approach, also involves its own set of assumptions. Here the main additional assumption is that the existing data represent some larger collection or *ensemble* of well-defined outcomes. These results have not yet been obtained as part of any experiment; they are, instead, a sort of theoretical context in which the existing results can be interpreted. The observed data are interpreted as a random selection of points from this larger ensemble. Frequentists prefer to assume that a particular hypothesis is true and then investigate how well the existing data agree with the (presumably true) hypothesis. This, at least, provides them with a testable hypothesis. If the agreement is good there is no reason to reject the hypothesis.

The frequentist view of probability quickly displaced the Bayesian view of probability, but in 1939 the British astronomer, geophysicist, and mathematician Harold Jeffreys (1891–1989) began to argue that the Bayesian approach had merit and that the frequentist view of probability had its own difficulties.

Jeffreys was born in a small town with the delicious name Fatfield in northeast England. He attended Rutherford College and Armstrong College, both of which are located in Newcastle-upon-Tyne. Jeffreys spent most of his working life at Saint John's College, Cambridge. He was an eclectic scientist, who made interesting contributions to several branches of science and mathematics. In mathematics he is probably best remembered for a well-received book on mathematical physics, *Methods of Mathematical Physics*, which was published in 1946, and for his *Theory of Probability*, published in 1939. By the time *Theory of Probability* was released the frequentists had (for the moment) won the debate. Their view of probability and its relationship to data had gained almost universal acceptance. In his book, however, Jeffreys criticized the frequentist approach and used Bayesian methods to examine many problems that formed a standard part of the education of any frequentist.

Initially, Jeffreys was a lone voice pointing to the possibility that there is, at least, something right about Bayesian probability, but when the book was published, it did not sway many minds. In retrospect, it is clear that Jeffreys fired the first shot in a resurgence of interest in Bayesian methods. In the intervening decades more and more mathematicians have become interested in this alternative understanding of what probability is. Nowadays the Internet is full of discussion groups that debate the merits of both sides.

7

THREE APPLICATIONS
OF THE THEORY OF
PROBABILITY

Discussions about the philosophy of probability and the importance of developing an axiomatic basis for the subject might
make probability appear irrelevant to the lives of most of us.
Discussions about the use of probability in estimating the number π or computing the mass of Saturn might make probability
appear to have only academic importance. But probability is now
one of the most used and useful branches in all of mathematics.
It also plays a vital role in many areas of society. In this section
we outline three contemporary applications of this mathematical
discipline.

Nuclear Reactor Safety

Commercial nuclear power reactors are highly complex machines.
Their purpose is to turn large quantities of heat energy into large
quantities of electrical energy. The electrical energy is then transmitted along power lines so that we, the consumers, can consume it.
The huge amounts of thermal energy produced and the large forces
necessary to convert that energy into electrical energy must be carefully controlled. Safe operation of a reactor requires a very thorough
knowledge of the way these power plants work. Nuclear reactors
are, perhaps, the most thoroughly analyzed machines ever built.
The goal of much of this analysis is to ensure that each nuclear plant
operates in the manner in which it was intended: Each plant is

Trojan Nuclear Power Plant. The theory of probability is an important tool in evaluating the safety of nuclear power plants. (Courtesy National Archives, College Park, Maryland)

designed to produce electrical power without endangering lives or unduly disrupting the environment. One of the analytical tools used to predict nuclear plant performance is probability theory.

When a reactor is producing electrical power much of the actual machinery in the plant is idle. These are the backup systems. They are supposed to be idle. They are designed to be brought on-line only when another system in the plant fails. Each system is individually tested and the data collected are used to evaluate the reliability of that system. Engineers can use these data to predict the probability of a fault within the system. Knowing the probability of failure of each individual system is not, however, enough to predict reliably how the plant will operate. The goal of collecting this information is to use it to develop a probabilistic model of the way the plant as a whole will work in the event of a failure of one of its components. To understand how this is done we can imagine each safety system in the plant as occupying a place on an organizational chart. At the top of the chart are the primary systems, those systems that should be working whenever the plant is producing power. Beneath each primary system is a treelike structure that shows how the functions of that primary system will be passed along to one or more backup systems should the primary system fail. Beneath the secondary systems we might also find tertiary systems that would be put on-line in the event of failure of the secondary systems.

The goal of the safety analyst is to imagine what failures, either mechanical or human, may occur in the operation of each system. These failures are played out in numerous scenarios in which the analyst computes how a failure in one system at one level would be distributed to one or more systems at the next level down. With this understanding of the safety architecture of the plant, a catastrophic failure would involve numerous individually unlikely failures working their way down the *event tree* from top to bottom. The analyst uses data about the individual systems, including the control system that transfers functions between the nodes of the event tree, to estimate the probability that a failure could work its way along the tree from top to bottom. These event trees are used to evaluate and compare different designs for possible future construction. They are also used to evaluate the safety of plants currently in operation. Decisions about when a particular nuclear plant is safe enough to operate are crucial to all of us. Probability theory provides one important technique by which these decisions are made.

Markov Chains and Information Theory

The theory of digital communication began shortly after World War II with the work of the engineer and mathematician Claude Shannon (1916–2001). In 1948, while at Bell Labs, he published a series of papers, *A Mathematical Theory of Communication*. His goal in these papers was to characterize mathematically the transmission of information. Doing this, of course, requires a mathematically acceptable definition of what information is. The definition has to be applicable to everything that we want to call a "message." Shannon liked concrete examples, and his paper has numerous "artificial languages," simple examples of messages consisting of strings of letters arranged so that each letter appears with a preassigned frequency. But his work is much broader than his examples indicate. Shannon's definition of information is really a statement about the amount of order or predictability in any message. It has nothing to do with the actual content, and today information theory is applied to problems in genetics and linguistics as well as digital communications.

According to Shannon the transmission of information requires (1) a source of information (the source generates a string of symbols, usually a sequence of numbers or letters), (2) a transmitter that encodes or changes the sequence generated by the source into a form suitable for transmission, (3) a channel along which the information is conveyed, and (4) a receiver that decodes the information conveyed across the channel and displays it for the recipient. Central to Shannon's model is the presence of noise in the channel, where *noise* means occasional, random changes in the information stream.

In Shannon's mathematical model the sequence of symbols that is generated at the source is a Markov process. In a long message the probability of the next symbol is determined by the order of the symbols just received. (The probability of moving from the set of received symbols to the next, new symbol is usually described as a type of Markov chain.)

Goldstone Deep Space Network is used to communicate with interplanetary space probes. (Courtesy of the National Aeronautics and Space Administration)

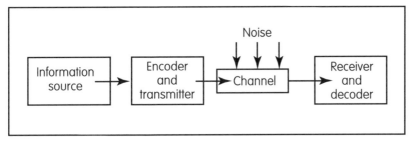

A model depicting how information is transmitted

Shannon's discovery of this definition of information allowed him to show that information *as he defined it* obeys certain laws that are in some ways analogous to those laws that describe the rate of change of other physical quantities such as mass, momentum, and energy. By employing the theory of probability, and especially the theory of Markov chains, he was able to show that information can be transmitted with extremely high accuracy even when the channel is noisy, provided that the information is correctly encoded at the transmitter. This was a surprising result since before Shannon's work, it was generally assumed that on a noisy channel parts of the transmitted message would inevitably be lost. Shannon's discovery led to the search for optimal error-correcting codes, codes that were as fast as theoretically possible and that still preserved the message in the presence of noise. Error-correcting codes are now routinely used throughout our society. They make it possible, for example, for the *Voyager* space probes, now located at the farthest reaches of our solar system, to continue to communicate with Earth successfully by using 23-watt radios. More generally, Shannon's discoveries are the foundation of all work in digital communication, because they made it possible to develop a successful mathematical model for the transmission and storage of information.

Smallpox in Modern Historical Times

Smallpox has long been a public health scourge. It had been a major source of mortality in the Eastern Hemisphere for thousands of years. It spread throughout North, Central, and South America with the arrival of European colonists, explorers, and conquerors. The disease is rapidly transmitted between individuals, and there has never been an effective treatment for those unfortunate enough to become infected. Generally, about one-third of all those who were infected died, although among Native Americans, in particular, the fatality rate was much higher. Throughout history a great deal of thought has been given to controlling and eliminating smallpox.

The technique of variolation, the dangerous but often effective technique of conferring resistance to smallpox, was of profound importance. (See the section on Daniel Bernoulli and Jean d'Alembert earlier in this volume for background on variolation.) One of the peculiar aspects of variolation is that it depends on the existence of individuals infected with smallpox so that the smallpox "matter" from the infected individual can be used to induce

Smallpox vaccination program, 1946, Jewell Ridge, Virginia (Courtesy National Archives, College Park, Maryland)

immunity in the healthy individual. The great breakthrough occurred with the work of the British surgeon and scientist Edward Jenner (1749–1823). Jenner was aware that a person who has been infected with cowpox, an illness that is not life-threatening, is thereafter protected from smallpox. He devised an experiment: He found a woman who was sick with cowpox. He removed some matter from a lesion on her finger and infected an eight-year-old boy. The boy contracted a mild fever and a small lesion. When the boy recovered, Jenner infected him with smallpox, but the boy did not become sick. Jenner had discovered a safe means to induce immunity to smallpox in a way that did not depend on the existence of other individuals infected with smallpox. Because protection is permanent and smallpox can be transferred only from person to person, Jenner's discovery set the stage for the eventual elimination of the disease.

Widespread inoculation wiped out smallpox in the United States in the late 1940s. Because smallpox was still present in other countries at that time there was still a danger that the disease could be reintroduced into the United States, and so for decades, at great cost, the United States continued the practice of widespread inoculation. The benefits of this public health policy were obvious: Smallpox is a scourge and the vaccine is extremely effective. There were also risks associated with this procedure, but at the time the risks were so small compared with the benefits that they received little attention. Remember: Millions of people lived in the United States, there was continual movement of people across the nation's borders, and for years not one person in the United States had died of smallpox. It is easy to see why the risks associated with the vaccine drew little attention. But the risks were present. In particular, there was a very low but predictable fatality rate associated with the vaccination.

As years passed without any additional cases, public health officials began to reassess the vaccination program. One new source of concern was the mortality rate associated with vaccination. Although the rate was quite low, it still exceeded the actual number of smallpox cases in the United States, which had been zero since late 1949. The tremendous success of the vaccination program introduced a new question: Was the vaccination program itself now causing more deaths than it was preventing?

A second source of concern to public health officials was the cost of the general vaccination program itself. Inoculating the entire population against smallpox was expensive. Critics began to ask whether this money might not be better spent elsewhere, curing or preventing other more immediate threats to the public health. The theory of probability was used to compare the risks to the public health of continuing the program of mass inoculation versus discontinuing it. Two questions were particularly prominent in the analysis: First, what was the probability of another outbreak of smallpox in the United States? Second, if an outbreak did occur, what was the probability that the disease could be quickly contained? Analysis of the available data showed that the risk of additional outbreaks was low and that the public health services could probably quickly contain any such outbreaks. Though the mathematics involved had advanced considerably, the questions were still reminiscent of Daniel Bernoulli's attempt to determine the efficacy of variolation centuries earlier and of d'Alembert's critique of Bernoulli's reasoning. In 1972 the United States discontinued its program of routine inoculation against the smallpox virus.

In 1977 the last naturally occurring case of smallpox was recorded. The world was smallpox-free. As a naturally occurring disease, smallpox had been destroyed. In 1980, amid much fanfare, the World Health Organization declared that the smallpox danger was over. There was not a single infected human being on the planet. Because smallpox can be contracted only from another infected individual, smallpox had been wiped out as a *naturally occurring* threat. That would have marked the end of the smallpox danger to humanity had not humans made a conscious decision to preserve the smallpox virus. Today, smallpox continues to exist in a few laboratories, where it is maintained for research purposes.

After the destruction of the World Trade Center in 2001, concern about bioterrorism increased. Public health officials began to contemplate a new possibility: After the elimination of smallpox as a public health threat, people might now deliberately reintroduce the disease into the general population as a weapon of war. Officials were once again faced with the same questions about the relative risk of smallpox attack versus the risk of the vaccine itself. Public health officials were faced with the decision of whether or not to reintroduce a program of mass vaccination that might cause more deaths than it prevented. (The vaccine will prevent no deaths if there is never an attempt to reintroduce the disease into the general population.) Once again we face the same sorts of questions that Bernoulli and d'Alembert contemplated centuries ago.

PART TWO

STATISTICS

INTRODUCTION

THE AGE OF INFORMATION

We are awash in a sea of data. Words, numbers, images, measurements, and sounds have been translated into trillions of strings of binary digits suitable for transmission, manipulation, and storage. At one time, generating large amounts of data involved a lot of work. The U.S. Census is a good example. Every 10 years census workers fan out across the nation collecting information on how many people live in the United States and where they make their home. For most of the history of the census, this information was collated and printed in thick, multivolume sets of books. These old books are not only a snapshot of the United States at a particular time in history; they are also a window into the way people used to work with large data sets. If we thumb through an old volume of the U.S. Census, we find page after page after page of tables. The size of the data set is numbing. It is difficult to see how a single individual could comprehend such a huge collection of numbers and facts or analyze them to reveal the patterns that they contain. Large data sets preceded the existence of the techniques necessary to understand them. To a certain extent they probably still do.

Our ideas of what constitutes a "large" data set have changed as well. Today, almost anyone can, over the course of a few months, generate a data set with a size similar to that of the U.S. Census by using a laptop computer. Simply attach a few sensors or measurement devices to a computer and make a measurement every fraction of a second and store the results. Such homegrown data sets may not be as valuable as census data, but their existence shows that our conception of what it means to collect a large data set has changed radically.

Collecting useful data is not easy, but it is just the first step. Data are the foundation of any science, but data alone are not enough. For large data sets, simply having access to the measurements provides little insight into what the measurements mean. Information, no matter how carefully collected, must *mean* something before it has value. Some of the questions we might ask of a data set are obvious. What is the largest value in the set? The smallest? What is the average value, and how spread out are the values about the average? Other questions, such as the existence of relationships among different classes of values, are not at all obvious. All such questions are important, however, because each question gives the investigator more insight into what the data are supposed to represent.

Statistics is a set of ideas and techniques that enable the user to collect data efficiently and then to discover what the data mean. Statistics is an applied discipline. In colleges and universities it is sometimes offered through a separate department from the department of mathematics. There is, nevertheless, a lot of mathematics in statistics. Research statisticians routinely work on difficult mathematical problems—problems whose solutions require real mathematical insight—and they have discovered a great deal of mathematics in the course of their research. The difference between mathematics and statistics is that statistics is not a purely deductive discipline. It involves art as well as science, individual judgment as well as careful, logical deductions.

Practical considerations motivated the development of the science of statistics. Statistics is used as an aid in decision making. It is used to control manufacturing processes and to measure the success of those processes. It is used to calculate premiums on insurance policies. It is used to identify criminals. In the health sciences, finding a new statistical relationship between two or more variables is considered ample grounds to write and publish yet another paper. Statistics is used to formulate economic policy and to make decisions about trading stocks and bonds. It would be difficult to find a branch of science, a medium or big business, or a governmental department that does not collect, analyze, and use statistics. It is the essential science.

8

THE BEGINNINGS
OF STATISTICS

In the winter of 1085 William the Conqueror, duke of Normandy and king of England, was in Gloucester, which is located by the Severn River in western England. He met with his advisers and out of this meeting arose the idea of obtaining a "description of England." The level of detail and the efficiency with which the data were obtained make the description one of the great accomplishments of the European Middle Ages. The description, which is preserved in two volumes, was known locally as the Domesday Book, the record against which one had no appeal. (*Domesday* was the Middle English spelling of *doomsday*.)

The data were collected and ordered geographically. The information consists of a long list of manors, their owners, the size of the property, the size of the arable property, the number of teams of oxen, other similar measurements, and a final estimate of the value of the property.

Groups of commissioners fanned out to the various counties. They set up a formal panel of inquiry in the principal town of each county. It was serious business. Representatives from each locality were called before the commissioners and asked a series of questions. All representatives had to swear to the truthfulness of their answers. The information was collected, ordered, and sent to Winchester, the capital of England at the time. Finally, the list was compiled and summarized into the text called the *Domesday Book*. The *Domesday Book* survives and is on public display at the Office of Public Records in London.

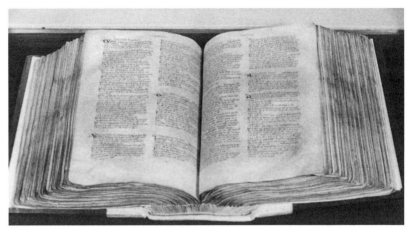

The Domesday Book *is still a subject of scholarly research more than 900 years after it was first compiled.* (Topham/The Image Works)

The *Domesday Book* is, essentially, a record of William's newly acquired holdings—he had conquered England in 1066—and it is generally thought that one reason for creating the record was its value as a tool for determining tax rates. The book itself, however, gives no indication of its intended use, and William died in 1087 before he could use the book for anything. Nevertheless, if a government official wanted to know how many teams of oxen, a common measure of wealth, were at a particular location in Nottinghamshire, he could find out by consulting the text. Insofar as anyone wanted answers to these extremely simple questions, the list was useful. Today it is tempting to summarize the list with statistical language—to calculate averages and correlations—to mine the data for the information they provide about life almost 1,000 years ago on a cold, backward fragment of land off the coast of the continent of Europe, but this is not how William and his contemporaries could have used the book. To be sure, this information would have been valuable to William. It could have assisted him in calculating tax rates and identifying which areas of the country were efficiently managed and which needed improvement. Most of the information that was so carefully collected by William was, however, unavailable to him. Without statistics there was no way to discover most of the information that lay concealed within the data—information that his commissioners had so carefully collected.

The *Domesday Book* is unusual because it was so well done. In its approach to information, however, it is typical: Much more information is collected than is revealed by the prestatistical analysis of the time.

The Beginning of Statistics

The history of statistics is unusual in that it begins with the work of a single person, the British shopkeeper and natural philosopher John Graunt (1620–74). Little is known of Graunt's early life. It is known that he was a successful businessman, a city councilman, an officer of a water company, and a member of the militia. He also had an inquiring mind. He reportedly rose early each day to spend some time studying before opening his shop. What his initial motivation was for these early morning study sessions is not known, but they eventually led him to consider human mortality. Many people think about life and death, of course, but perhaps for the first time in history, Graunt sought systematic, quantitative information on the subject. He turned an analytic eye on who lived, who died, and why, and what is more, he was the first person in history to find answers.

His sources were the Bills of Mortality. These were lists of christenings and deaths that a clerk in each parish would compile weekly. In addition to simply noting that someone died, each entry also listed a cause of death. The practice of issuing Bills of Mortality had begun several decades before Graunt began to consider them. Apparently, the practice had been initiated in 1592 in response to the high mortality rate caused by the plague. At first the bills were issued sporadically, but in 1603 their issue became standard, and thereafter each parish submitted a bill each week. Graunt analyzed the data from 1604 to 1661. His sole publication, *Natural and Political Observations Mentioned in a following Index, and made upon the Bills of Mortality*, is an extraordinarily creative piece of research.

In *Observations* he describes the criteria he used to analyze a problem, and he lists some of the facts that he discovered through a careful study of the bills.

What possessed Graunt to labor over 57 years of Bills of Mortality? What was the reason for what he called his "buzzling and groping"? Graunt lists several reasons. His haughtiest, and in some ways his most personally revealing, reason is that if one cannot understand the reason for his questions then one is probably unfit to ask them. Another personally revealing answer is

CAUSES	VICTIMS	CAUSES	VICTIMS
Convulsion	241	Piles	1
Cut of the Stone	5	Plague	8
Dead in the Street,		Planet	13
and starved	6	Pleurisie, and Spleen	36
Dropsie, and Swelling	267	Purples, and spotted	
Drowned	34	Feaver	38
Executed, and prest		Quinsie	7
to death	18	Rising of the Lights	98
Falling Sickness	7	Sciatica	1
Fever	1108	Scurvey, and Itch	9
Fistula	13	Suddenly	62
Flocks, and Small Pox	531	Surfet	86
French Pox	12	Swine Pox	6
Gangrene	5	Teeth	470
Gout	4	Thrush, and Sore	
Grief	11	mouth	40
Jaundies	43	Tympany	13
Jawfaln	8	Tissick	34
Impostume	74	Vomiting	1
Kil'd by several accidents	46	Worms	27

Some examples from the list of causes of mortality for the year 1632, compiled from the London Bills of Mortality by John Graunt. In a period when record keeping was increasing and societies had little understanding of the nature of disease, mathematicians began to turn to statistics for insight into health risks.

that he enjoys deducing new facts from a study of what we would call the "raw data" of the Bills of Mortality. He gives other, more scientific reasons for his fascination with the bills. These show that he fully understood the importance of what he had done. Graunt had learned how to use data to estimate risk.

The evaluation of risk is not something that can be done by simply glancing through the Bills of Mortality. This, he says, was what most people did with them, looking for odd facts, unusual deaths, and so forth. Instead, the evaluation of risk can be done only by collating the data and performing the necessary computations. This could not have been easy. Over a half-century of data collected from numerous parishes yielded a record of 229,250 deaths. In addition to the total number of deaths, he made a list of all the ways people died. There were plague, smallpox, "bit with a mad dog," measles, murder, execution, "dead in the street and starved," suicide, stillbirth, drowning, "burnt and scalded," and a host of other causes. Each cause is carefully listed, and many are analyzed. He discusses, for example, whether a cough is the correct "cause" of death of a 75-year-old man or whether old age is not a better diagnosis for anyone in that age group. All of this goes to the accuracy of diagnoses and the reliability of the data. It is in some ways a very modern analysis. After he describes his thinking on these matters, he begins his search for the truth.

Murder, then as now, was always a popular topic of conversation. Any murder diverts our attention and causes us to assess our own safety. How great a danger was death by murder to the people of Graunt's place and time? To answer that question he turned to the Bills of Mortality. He computed that only 86 of the total of 229,250 died of murder. Murder, he shows, is not a significant cause of mortality. In Graunt's hands the Bills of Mortality also reveal that the popular wisdom that plagues accompany the beginning of the reign of a king is false. Each short paragraph of his analysis uses data to dismiss certain common fallacies and to discover new truths. All of this information, he tells us, will enable his contemporaries to "better understand the hazard they are in."

Graunt also examines the birth rates and discovers a fact that would fascinate many subsequent generations of scientists: More male babies are born than female. He writes at length about this discovery. Males, he tells us, are more likely to die in war, die on the open seas, face the death penalty, and so on, but because males have a higher birthrate the numbers of adult women and adult men are roughly equal.

It should be emphasized that all of this information was already in the Bills of Mortality. Graunt's great insight was to analyze the bills systematically, to extract information from data, and to use mathematics to reveal relationships among numbers. This was new.

Unlike so many new ideas, Graunt's observations were immediately recognized as highly original and valuable. Graunt was a shopkeeper, not a scholar, and his research was of a type that could not be easily classified. The Royal Society, the most prestigious scientific organization in his country, might not have been disposed to admit him as a member, but King Charles II himself interceded on Graunt's behalf. Not only did Graunt become a member, but also the society was instructed to admit "any more such Tradesmen" and "without any more ado." The *Observations* was published in several editions and influenced research in England and on the Continent. This was the beginning of a new era, an era in which people thought statistically. Nowadays we learn to think statistically as children, but it has not always been so. John Graunt was the first.

Edmund Halley

The name of the British mathematician and scientist Edmund Halley (1656–1742) is permanently linked with Halley's comet. Halley was not the first person to observe this comet, but he was the first to predict its reappearance. He had studied records of a number of comets and formed the opinion that one of the comets in the record was periodic. In 1705 he published his calculations that indicated that the comet would return in 1758. Although Halley did not live long enough to see his prediction verified, the comet was again sighted in 1758 just as he had calculated. His name has been associated with the comet ever since. An adventurous spirit, Halley did much more than predict the return of a comet.

Halley was born into a prosperous family at a time when society's interest in science and mathematics was very high. There was, for example, a demand from many quarters for high-quality astronomical observations and mathematical models that would enable ships at sea to determine their position better. British ships were now sailing around the globe, and the ability to establish one's

Edmund Halley. Among his many accomplishments he helped establish the science of statistics. (Library of Congress, Prints and Photographs Division)

position from astronomical observations had become a matter of some urgency. All of this influenced, or perhaps coincided with, young Edmund Halley's interests. He was fascinated with mathematics and astronomy from an early age, and his family was in a position to outfit him with a number of high-quality astronomical instruments. He became quite proficient in using them, and when he arrived at Queen's College, Oxford, to begin his studies, he had enough equipment to establish his own observatory. He even brought his own 24-foot (7.3-m) telescope to college.

While at Queen's College, Halley visited the Royal Greenwich Observatory, a place that occupied an important position in English scientific life. He met the head of the observatory, John Flamsteed, who was an important scientific figure of the time. Flamsteed was involved in making precise measurements of the position of all stars visible from Greenwich, England. Halley soon embarked on his own version of the project: He left Queen's College without graduating and took a number of high-quality astronomical instruments to the volcanic island of Saint Helena, a British possession located about 1,000 miles off the coast of Africa in the South Atlantic Ocean. Once there, he established his own temporary observatory on the island.

Halley chose Saint Helena because it is located well south of the equator. The stars visible from Saint Helena are different from the stars visible from Greenwich. His goal was to make a star chart for the Southern Hemisphere to complement the one Flamsteed was engaged in making for the Northern Hemisphere. Although his work was hampered by cloudy nights, Halley succeeded in making accurate measurements of more than 300 stars. Halley's work was exemplary, and he was later awarded a master's degree from Queen's College at the behest of King Charles II.

Working with large data sets was something for which Halley had a particular aptitude. He had a practical and theoretical interest in the winds and the oceans. (In addition to his trip to Saint Helena he later took a second, more dangerous trip to the Southern Ocean and wrote a beautiful description of the huge icebergs he encountered there.) Another of his big projects involved collecting as many meteorological data as he could. He used the data to create a map of the world's oceans showing the directions of the prevailing winds. This was the first map of its type to be published, and the information it contained was of interest to every ship's captain of the time.

The data analysis of most interest to a history of statistics, however, was Halley's paper on mortality rates in the city of Breslau. (The city of Breslau is today called Wrocław and is located in western Poland.) Halley was aware of the work of Graunt, but his interests were more specific than those of Graunt, whose paper was a very broad inquiry. Halley wanted insight into life expectancy. The phrase *life expectancy* usually conjures up images of cradle-to-grave average life spans, along the lines of "How long can a baby born today be expected to live?" But Halley's questions were considerably more detailed. He wanted to know, for example, the probability that a 40-year-old man would live seven additional years. He examined this and several related problems. For example: For a randomly chosen individual of age n years, where n represents any given age, he wanted to know the probability that this randomly chosen individual would live one additional year. One more example: Given a group of individuals, all of whom are the same age, in how many years will only half of them remain alive? These types of questions are much more detailed than those considered by Graunt. They are the types of questions that must be answered by insurance companies in order to set life insurance premiums rationally. Today, the individuals who search for the answers to these and related questions are called actuaries, and the branch of science in which they work is called actuarial science. Halley's paper, "An Estimate of the Degrees of the Mortality of Mankind, Drawn from Curious Tables of the Births and Funerals at the City of Breslaw; with an Attempt to Ascertain the Price of Annuities upon Lives," is generally recognized as the first serious attempt at actuarial science.

For the application Halley had in mind, the Bills of Mortality that had been collected by Graunt had a serious shortcoming, or at least Halley suspected them of having a shortcoming. The problem was that London's population was too mobile. Halley had no way of knowing who was moving in or out or how the continual migration was changing the population. London was growing, but Graunt's Bills of Mortality showed that deaths were more common than births during this time. This could happen only if there were an influx of people from the countryside. Without more information about who was moving in and out it was difficult to make reliable deductions from these data. Halley decided to search for a large city that kept good records and also had a stable population. This meant that he needed to use a population who, for the most part, died near where they were born. He found that the city of Breslau satisfied these conditions.

As a matter of policy the city of Breslau compiled monthly bills of mortality that recorded several facts about each individual listed in the bill. Of special interest to Halley were the individual's age at the time of death and the date of death. Halley had access to records for five consecutive years (1687–91). These records were carefully compiled, but, as in the London Bills of Mortality and other similar records, most of the information that these bills contained was hidden from view because it had not been subjected

BRESLAU MORTALITY TABLE

AGE	LIVING	AGE	LIVING	AGE	LIVING	AGE	LIVING	AGE	LIVING
1	1,000	18	610	35	490	52	324	69	152
2	855	19	604	36	481	53	313	70	142
3	799	20	598	37	472	54	302	71	131
4	760	21	592	38	463	55	292	72	120
5	732	22	586	39	454	56	282	73	109
6	710	23	580	40	445	57	272	74	98
7	692	24	574	41	436	58	262	75	88
8	680	25	567	42	427	59	262	76	78
9	670	26	560	43	417	60	242	77	68
10	661	27	553	44	407	61	232	78	58
11	653	28	546	45	397	62	222	79	49
12	646	29	539	46	387	63	212	80	41
13	640	30	531	47	377	64	202	81	34
14	632	31	523	48	367	65	192	82	28
15	628	32	515	49	357	66	182	83	23
16	622	33	507	50	346	67	172	84	19
17	616	34	499	51	335	68	162	*	*

(*de Moivre's* The Doctrine of Chances)

to a statistical analysis. Halley uncovered quite a bit of information over the course of his analysis. He did this by constructing a table that lists the number of people of each age who were alive in the city at a given time. As the numbers in the age column increase, the numbers of people who are that age decrease. It is from this table that he drew his deductions. His analysis is an interesting insight into life in Breslau and probably much of Europe at this time. The following are some of the facts that Halley uncovered:

1. Breslau had a birthrate of 1,238 births per year and a death rate of 1,174 per year. Halley discovered that of those born, 348 infants died in their first year. (In more modern terminology, this represents an approximately 28 percent first-year mortality rate.)

2. Of the 1,238 born each year, on average 692 lived to see their seventh birthday. (This is a mortality rate of approximately 44 percent.)

3. The mortality rate can be analyzed by age. Halley divides the ages into different groupings and calculates the mortality rate for each. For example, for people between the ages of nine and 25 the death rate is roughly 1 percent per year, a rate, he remarks, that roughly coincides with that in London. He continues his calculations until "there be none left to die."*

4. In Breslau, population 34,000, it was possible to raise an army of 9/34 of the total population, or 9,000 men, where the population of men suitable for fighting consists of males between the ages of 18 and 56.

5. Halley also demonstrates how to compute the odds that an individual of any age will live an additional year, or, for that matter, to any proposed age. He uses the example of a man of age 40. His method is straightforward: He notes how many individual men are alive at age 40 (445) and how many are alive at age 47 (377). The conclusion is that during this time (assuming no migration into or out of the city) 68 died. Dividing 377 by 68 shows that at age 40, an individual has a roughly 11 to 2 chance of surviving until age 47. He also considers the following related, but more general problem: Given an age—for purposes of illustration he chooses age 30—compute to what year the individual of age 30 has a 50/50 chance of surviving. In Halley's table, individuals of age 30 have a 50/50 chance of surviving to a time older than age 57 but less than age 58.

Halley then goes on to make other deductions from the data, including deductions relevant to calculating insurance rates. At this point the mathematics becomes somewhat more complicated, and he provides geometrical proofs that his reasoning is sound. He closes in a more philosophical vein. He points out that although the people of his era (as of ours) often talk about the shortness of

INSURANCE

Historically, the insurance industry has been one of the most important users of statistics as well as a source of innovation in statistical techniques. What, then, is insurance?

Each of us makes plans for the future. Experience shows that those plans do not always work out. This is not always bad. Sometimes our plans are changed for the better. Sometimes, of course, they change for the worse. When they change for the worse, the losses we suffer can take a variety of forms. A change of plan can mean a loss of time, confidence, property, opportunity, or even life itself. Insurance is a method for compensating the insured for *financial* losses. The key behind the idea of insurance is that in order for an item to be insurable, it must have a well-defined cash value. Although there have been companies that have attempted to insure against, for example, "pain and suffering," historically

Flood damage along the Kansas River, 1903. Insurers expect the occasional disaster. This is why they are careful to distribute their risk. (Photograph by Sean Linehan, courtesy of National Oceanic and Atmospheric Administration/Department of Commerce)

life and how wronged they feel when they or someone they know fails to achieve old age, they have little sense of how long one can really expect to live. The data indicate that the average life span,

these companies have found it difficult to arrive at a reasonable cash value for a given amount of pain and suffering.

In order to insure an item, the insurer must have enough information on the item to estimate the probability that the item will be lost. (Here *item* could mean a possession, such as a house, or even one's life.) Insurers *expect* to pay for the occasional lost item. What they depend upon is that the losses they suffer will "balance out" in the sense that the insurers will make more money in premiums than they will lose in payments. This is where mortality tables like those studied by Graunt and Halley, as well as other conceptually similar sources of information, become important. From mortality tables, for example, life insurance companies attempt to calculate the probability that a randomly chosen individual will live to a certain age. The premiums are then calculated, in part, on the basis of the information in the tables. It is essentially a bet: If the insured lives to the specified age, then the life insurance company has earned a profit and the insured loses the money spent on premiums. If, however, the insured dies early, then the life insurance company suffers a loss and the insured—or at least the insured's beneficiary—collects the money. A great deal of effort has gone into computing premiums that are profitable to the insurer, and it is rare that a life insurance company fails to make its yearly profit.

In themselves, the tables do not contain enough information to enable the insurer to set rates. Other criteria have to be satisfied. Principal among these is the condition of randomness. No insurer will pay on a nonrandom loss. One cannot insure one's house against fire and then proceed to burn it down. Another general criterion is that the pool of individuals must be sufficiently dispersed so that they cannot all suffer a simultaneous loss. Many houses in New Orleans, Louisiana, have flood insurance, for example, but no commercial insurer specializes in providing flood insurance to the residents of New Orleans, since a single flood would bankrupt the insurer.

The science of insurance is called actuarial science, and it is a highly developed specialty. Actuaries are thoroughly versed in the mathematics of probability and statistics. In addition, they are familiar with concepts involved in pensions, annuities, and the general business of insurance. It is an oft-repeated expression that nothing is less certain than life and that nothing is more certain than the profits of a life insurance company.

the time at which half of those born have died, in Breslau at this time was just 17 years of age.

Halley's analysis of the Breslau data was his main contribution to the development of statistics, but even taken together with his other accomplishments in science, it is an inadequate measure of what he accomplished. Halley contributed to science and mathematics in other, less obvious ways. As many mathematicians profiled in this series had, Halley had an unusual facility with languages. During his lifetime some of the works of Apollonius were known only in Arabic translation. (Islamic mathematicians, in addition to producing a great body of work in the field of algebra, had produced a number of Arabic translations of important Greek and Hindu texts.) Halley learned Arabic in order to translate some of the work of Apollonius, and this he did.

Finally, Halley's name is connected with that of the British physicist and mathematician Isaac Newton (1643–1727). Halley had approached Newton to discuss problems relating to calculating the orbits of the planets, a problem that had, to his knowledge, not yet been solved. Newton, however, had already calculated that planets must, under the influence of gravity, follow elliptical orbits, but he had kept the discovery to himself for many years. When Halley learned of Newton's calculations, he immediately recognized the importance of the discovery and convinced Newton to publish the work. The result was *Philosophiae Naturalis Principia Mathematica* (Mathematical principles of natural philosophy), one of the most influential works in the history of science. It was Halley, not Newton, who oversaw publication. Halley wrote the introduction, proofread the manuscript, and—though he was practically broke at the time—paid for publication.

The mathematics that John Graunt and Edmund Halley used in their analysis of bills of mortality is, for the most part, exceedingly simple. Simple math is, in fact, characteristic of a lot of basic statistics. There are no new mathematical techniques in either Graunt's or Halley's papers. Their analyses consist, after all, of basic arithmetic, easily solved not only in our time but in Graunt's and Halley's as well. Furthermore, Halley was an excellent mathematician, so there can be little doubt that he, at least, found all of these calculations trivial. (Less is known about Graunt's abilities, simply because less is known about Graunt in general.)

What is new in both these papers is that the authors are extracting new statistical relationships from their data sets. They are discovering new facts about life and death in the cities of London and Breslau, and *these are some of the first instances of anyone's thinking of performing this type of statistical analysis.* Graunt's analysis is, in fact, *the* first instance of statistical analysis.

Statistics enable the user to extract meaning from data. Numbers, especially large collections of numbers, are usually not informative in themselves. The statistician's goal is to reveal the information that is contained in the numbers. Without statistical analysis collecting data is hardly worth the effort, but carefully collected data can, in the hands of a skillful statistician, reveal many new facts and insights. The works of Graunt and Halley are two of the most significant analyses in the early history of statistics.

9

DATA ANALYSIS AND THE PROBLEM OF PRECISION

John Graunt and Edmund Halley drew a number of insightful conclusions from their analyses of bills of mortality, but one problem they did not consider in a systematic way was the problem of precision. Graunt and Halley mentioned it, but the problems that they studied did not lend themselves to a rigorous mathematical discussion of precision. The problem of drawing precise conclusions from numerical data was first treated in the early 19th century by the French mathematician Adrien-Marie Legendre (or Le Gendre; 1752–1833).

Little is known of Legendre's early life. He would have had it no other way. He wanted to let his mathematical work speak for itself, and he had no interest at all in sharing personal details. To this day it is not clear where he was born. Some accounts indicate he was born in Paris; other cite Toulouse as his birthplace. It is certain that he was born into a wealthy family and that he grew up in Paris. He received an excellent education in mathematics and physics at Collége Mazarin in Paris.

He worked as an academic at the École Militaire and the École Normale, two distinguished institutions. As were those of most French mathematicians and, indeed, many French citizens of the time, Legendre's life was adversely impacted by the political chaos of the French Revolution (1789–99) and its aftermath. In Legendre's case, he lost his fortune. He eventually settled his financial affairs and lived frugally on his salary. As an old man he lost his position in a political dispute and lived the brief remainder of his life in poverty.

Legendre made many significant contributions to higher mathematics. He enjoyed problems in mathematical physics, the mathematical analysis of equations that arise in physics. In the course of studying the equations that describe how bodies interact gravitationally, he invented what are now called Legendre functions. In addition to his more advanced work, he authored a famous textbook on elementary geometry, *Éléments de géométrie*. This book was a reexamination of the ancient Greek text *Elements*, by Euclid of Alexandria, the most famous and long-lived textbook in history. Legendre simplified the presentation, added new results, and created what was, for Legendre's time, a much better textbook. His book became the standard text on Euclidean geometry throughout much of Europe and the United States for about 100 years.

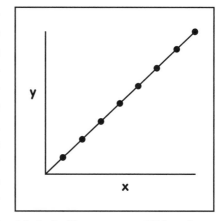

Method of least squares. This technique enables one to obtain the best linear approximation to the data at hand.

In middle age Legendre began to consider the problem of predicting the orbit of a comet from measurements, and this is where he made his contribution to statistics. His paper on this topic is called "Nouvelles méthodes pour la détermination des orbites des comètes" (New methods for the determination of comet orbits). The problem that Legendre wanted to solve is related to making the best use of measurements. He was faced with a collection of measurements, and he wanted to use them to determine the orbit of a comet. It might seem to be a straightforward problem: Make the required number of measurements and then perform the necessary computations. The problem is that every measurement contains some error, and as measurements accumulate so do errors. Minimizing the number of measurements is, however, no answer to the problem of uncertainty: With only a small set of observations, it is more difficult to estimate the size of the errors present in the data.

The solution to the problem of too few measurements is, of course, to make many more measurements. The more measurements one makes, the more information one has about the size of the error. It may seem, therefore, that having more information is always good, but before Legendre's work large numbers of measurements presented their own problem. There was no rational way to make use of large numbers of measurements. What Legendre did was to find a way of using the entire collection of measurements to *compute* a set of values that were optimal in the sense that (loosely speaking) the computed values minimized the amount of variation present in the data set. Usually these computed values are different from all of the measured ones. That does not matter. What is important is that Legendre's method yields a set of values that makes the best use of the data at hand. His method for doing this is now called the method of least squares.

The value of Legendre's discovery was immediately recognized. The math involved in implementing the method of least squares is not especially difficult, and this, too, is important. Not every scientist has a strong mathematical background, but all scientists who work with measurements—that is, the great majority of all scientists—can benefit from procedures that enable them to make the best use of the data. As a consequence Legendre's book on cometary orbits was reprinted a number of times, and his ideas on the method of least squares quickly spread throughout the scientific community.

In 1809 the German mathematician and physicist Carl Friedrich Gauss (1777–1855) published a paper written in Latin, "Theoria Motus Corporum in Sectionibus Conicus Solem Ambientium" (Motion of the heavenly bodies moving about the Sun in conic sections). This paper, as did that of Legendre, analyzed the problem of how to make best use of a series of measurements to predict the orbital path of a celestial object. (In this case, it was an asteroid rather than a comet.) As Legendre's paper did, this paper described the method of least squares, and, as Legendre had, this author also claimed to have invented the method.

Carl Friedrich Gauss, one of the great mathematicians of the 19th century, is the other originator of the method of least squares. Gauss was born in Brunswick and demonstrated his talent for mathematics at an early age. Fortunately, his abilities were quickly recognized, and he was awarded a stipend from the duke of Brunswick. In secondary school, Gauss studied mathematics and ancient languages.

The stipend continued as Gauss studied mathematics at the University of Göttingen (1795–99) and even after he received his doctorate from the university at Helmstedt in 1799. After graduation he could have found work on the strength of his thesis—for his degree he proved a remarkable result now called the fundamental theorem of algebra—but he did not look for employment because he was still receiving his stipend. Eventually, the duke of Brunswick died, and the support that Gauss had been receiving was discontinued. It was then that Gauss found a position at the University of Göttingen, where he taught an occasional mathematics course and was head of the university's astronomical observatory. He remained in this position for the rest of his life.

Gauss believed it was very important to publish only papers that were very polished. It was not uncommon for him to wait years to publish a result. Later, when another mathematician would publish something new and creative, Gauss would claim that he had already discovered the result. This type of behavior by most people is quickly dismissed, but it was often shown that Gauss had, in fact, originated the idea as he claimed. Gauss, however, was hard-pressed to prove that he had had the idea before Legendre, who vigorously objected to Gauss's claiming priority over this important idea. There are two points of which we can be sure: First, Legendre published first and so influenced the course of statistics. Second, Gauss's tremendous personal prestige has ensured that some historical accounts attribute the method of least squares to him.

Legendre and Gauss were two of the foremost mathematicians and scientists of their time, and both were at home with the most advanced mathematics of their age. Mathematically, however, the

THE MISUSE OF STATISTICS

When a set of measurements is analyzed statistically, we can learn a great deal about the set of measurements. This is true for every set of measurements, because every set of measurements has a statistical structure. It has a mean or average value and a variance, which is a measure of how widely scattered the measurements are about the mean. If the data points can be represented as points on a plane we can construct a line that is the "best fit" for those points, in the sense that the points are closer to the best-fit line than they are to any other line on the plane. If the data points can be represented as points in a three-dimensional space we can—subject to certain restrictions—find a surface that is the best fit for the given points. All these concepts and techniques—among many others—help us understand the structure of the data set.

Usually, however, it is not the data set as such in which scientists have an interest. Measurements are collected and analyzed because they are supposed to be representative of an even larger set of measurements that for reasons of time or money were not collected. This larger, unmeasured population is what the researcher really wants to understand. The smaller data set is supposed to give the researcher insight into the larger population. It is in this conceptual jump from the smaller collection of measurements to the larger, purely theoretical population that researchers often err.

Mistakes arise in a variety of ways. For instance, during the 19th century a number of studies of the head size of criminals were made. The scientists of the time were searching for a difference in head size or shape that would indicate a predisposition to a life of crime. The head sizes of a group of "criminals"—which usually meant a collection

method of least squares is not among these advanced techniques. It is typical of much of the statistics of the time: It is conceptually important, but mathematically simple. It is extremely useful and, as for so many statistical innovations, its usefulness was recognized immediately. It soon found its way into every branch of science that employed statistics. The method of least squares is still an important part of any introductory statistics course today.

of prisoners—were compared with the head sizes of a group of individuals who were not incarcerated. Of course, researchers found differences. Given any two sets of independent measurements, comparisons will always yield differences between them. The question is not whether there will be differences, but whether the differences are significant. Definitive conclusions notwithstanding, the study of the relationship between head size and criminal activity has since been largely abandoned.

Another area of statistical abuse has been in the area of intelligence testing. During the last 100 years, intelligence testing has become a small industry. Researchers have administered intelligence tests to groups of students and, on the basis of the tests, decided that the students who took the tests are less intelligent than, more intelligent than, or as intelligent as the general population. These conclusions were sometimes motivated by the statistical assumption that the students were a good cross section of the general population, an assumption that is a common source of error. Sometimes, for example, the students were nonnative speakers of English, the language in which the test was administered. No one can be expected to score well on a test if he or she cannot read the language in which the test is written. Clearly, the test was designed for a population to which the students did not belong. The results of the test, though they may have had a significant impact on the student's educational opportunities, could not reflect the ability of the students who were not fluent in English. There have been numerous instances of this type of error.

Any statistical conclusions that are drawn from an invalid statistical hypothesis such as this are suspect. Unfortunately, designing a standardized test that does not do a disservice to some part of the population has proved to be extremely difficult.

10

THE BIRTH OF
MODERN STATISTICS

The analysis of measurements in the physical sciences contributed to the early development of statistics. Physical scientists such as Laplace and Legendre required statistics as a tool to analyze the measurements they made. But physics and chemistry in the 19th century required only a modest amount of statistics to analyze the data that scientists collected. In part, this was due to the nature of the experiments and observations: As a rule, experiments in the physical sciences are easier to control. As a consequence they tend to yield data sets that are simpler to analyze.

Statistics took a huge leap forward as researchers turned their attention to the life sciences. In the life sciences and social sciences, randomness is an integral part of the subject. Carefully controlled experiments are often not possible; complex data sets are unavoidable. This difference between the physical sciences—physics and chemistry—and the life sciences is worth examining further to understand why statistics, an applied subject, developed largely in response to problems in the social sciences, the manufacturing sector, and especially the life sciences.

Physical science in the 19th century was founded on conservation laws: conservation of mass, conservation of momentum, and conservation of energy. These laws are expressed as equations—statements of equality between two quantities. No scientist would say that energy is "probably" conserved or that mass is "usually" conserved during a chemical reaction. These quantities are always conserved. By contrast, the great discovery in the life sciences of the 19th century, the theory of evolution, is a statistical theory: Certain gene frequencies are *more likely* to increase from one generation to the next than are other gene frequencies. Chance events play an integral part in which genes are passed from one generation to the next. But even in conditions in which chance events are less important, ecological systems are so complicated that identifying cause-and-effect relationships is often a hopeless task. Measurements made of such complex systems cannot be understood without a great deal of statistical insight.

The scientists who contributed most to the development of the science of statistics were—characteristically—multifaceted, multitalented, and fractious. They were very much involved in statistical applications. For them, statistics was not an isolated discipline; it was a tool to be used in the study of other subjects. These scientists, who were often experts in fields other than statistics, developed statistics in parallel with their studies in the life sciences, social sciences, manufacturing processes, and other disciplines.

Karl Pearson

One of the most historically important and creative statisticians was the British mathematician, lawyer, writer, art historian, and social activist Karl Pearson (1857–1936).

Karl Pearson was educated at King's College, Cambridge, where he studied mathematics. He also spent some time in Germany as a student at the universities at Heidelberg and Berlin. It was apparently on this trip abroad that his horizons expanded. While in Germany he studied math and physics, but he also studied philosophy and German literature. He became interested in the writings of Karl Marx, one of the founders of communist philosophy. On his return to Great Britain he studied law for a few years, but although he was qualified, he never showed much interest in a career in law. He did, however, begin to publish books. His first books, *The New Werther* and *The Trinity: A Nineteenth Century Passion Play*, were criticisms of orthodox Christianity. Because he is remembered as a statistician, one would think that he soon began to publish papers on statistics, but that occurred still later. In fact, one of his first research papers was on a topic in art history.

Pearson gave a lot of thought to issues of social justice. In an age when women's rights were very much restricted, Pearson argued for greater rights for women. He also advocated more openness in discussions of sex and sexuality—at the time, a very radical idea to advocate. Nor did his fascination with Marxism diminish. He wrote about socialism and prominent socialists wrote about him. Vladimir Ilyich Lenin, the founder of the Soviet Union and one of the most influential socialists of all time, followed Pearson's writings and wrote complimentary remarks about Pearson's ideas.

Pearson was hired to teach applied mathematics at University College, London, but his mathematical output at this time was not great. He did, however, continue to pursue his other interests. The turning point in his life as a mathematician occurred when he was hired as professor of geometry at Gresham College. In this capacity he became interested in graphical representations of statistical data. One problem of particular interest to him was the problem of curve fitting. To understand the idea, imagine a set of data relating two quantities. The quantities may, for example, be the height and weight of a collection of people, or the temperature and pressure of air inside a closed container as the volume of the container is varied. Each pair of measurements can be represented as an ordered pair of numbers, and each ordered pair of numbers can be represented as a point on a plane. If we now graph the set of all such points, they will form a pattern on the plane. This is the type of geometric pattern Pearson sought to analyze.

No set of measurements is complete. Even when the researcher collects many measurements, the assumption is that there are many more such measurements that could have been collected but were not. Consequently, the pattern of data that appears on the plane, however large and detailed it may at first appear, is assumed to be a modest representation of a much larger, more detailed pattern that could have been collected but, for whatever reason, was not. It is this larger, hypothetical set of points that the researcher wants to understand, because it is from this larger, "parent" set that all measurements are presumed to be drawn.

One way to understand a set of two-dimensional measurements is to draw a curve to represent the set. The curve can reveal a more precise mathematical relationship between the two measured quantities. After the curve that best represents the "few" points that have been plotted has been found, the researcher uses the curve to discover relationships between the two quantities represented by the data points. Finding the curve that best represents the data points is the problem of curve fitting.

For theoretical or practical reasons the researcher is always restricted to a particular family of curves. These curves cannot possibly pass through every data point that has been plotted on the plane. In fact, the curve that the researcher finally chooses may very well miss every point that has been plotted. Whether or not the curve hits any points is not in itself important. The goal is to find a curve that, when properly measured, is "close" to as many points as possible. (See the accompanying diagram for an illustration of this idea.) This means that the researcher must have a rigorous definition of what it means to say that the *set* of all points

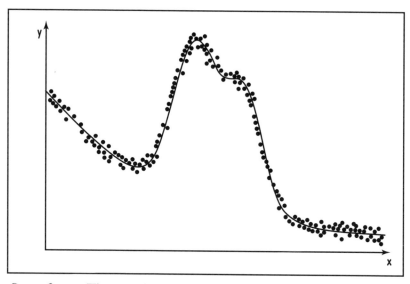

Curve fitting. The goal of the researcher is to choose, from a predetermined family of curves, the curve that best represents the data at hand.

off the curve is close to the curve. Furthermore, the researcher must show that the curve chosen is the best of all possible curves in the family of curves under consideration. These were new problems during Pearson's time, and in his position as professor of geometry, Pearson was kept busy seeking their solution. His ideas on this topic constitute one of his early contributions to statistics.

While at Gresham College, Pearson had contact with the British academic Walter Frank Raphael Weldon (1860–1906), and they became friends. Weldon also had very wide ranging interests, one of which was Charles Darwin's theory of evolution. Darwin proposed that species could change over time. According to Darwin these changes were hereditary. In particular, that means that the changes did not occur to individuals; rather, they occurred *between* generations, as the hereditary material was transmitted from one generation to the next. Darwin called this process of change *natural selection,* the tendency of some individuals to be more successful than others in passing on their hereditary traits to the next generation.

Weldon recognized that natural selection depended on the existence of many small differences between individuals of the same species. Previously, these small differences had been largely ignored as naturalists sought to develop an orderly system to categorize plants and animals by species. For purposes of categorization small individual differences *must* be ignored; to take them into account would make any classification scheme too complicated to use. For purposes of evolution, however, small individual differences, and their effect on the bearer's ability to transmit these differences to the next generation, were the key to understanding the process.

Developing a coherent way of describing a large set of small differences was precisely what Weldon needed, but Weldon was no mathematician. This was Pearson's contribution: He worked to make sense out of the great wealth of variation that is present among the individuals of most species. The ideas with which Weldon was struggling were exactly the ones that allowed Pearson to test his ideas—the ideas that he had been developing as a professor of geometry. Pearson and Weldon began to collaborate.

In his attempt to make sense of large and complicated data sets Pearson invented several concepts of modern statistics. In fact, Pearson's ideas *depended* on the data sets being large. Large data sets may seem more difficult to analyze than small ones, but in statistics the opposite is usually true. Many statistical techniques depend on the fact that the data set is large. Some of these techniques were pioneered by Pearson.

Notice that Pearson's ideas about curve fitting were purely mathematical. Some of his contemporaries even criticized Pearson for being too mathematical. For Pearson, each curve represented a mathematical relationship between two variables. It reflected the distribution of points in a plane. The curve did not necessarily depend on any deeper insight into the problem from which the data were obtained. Consequently, Pearson sought and found a test to measure how probable it was that the curve was a reasonable statement about the relationship between the two variables. The idea is to assume that the curve is actually a good representation for a carefully defined, larger set of points, and then to imagine randomly drawing a representative sample from this larger set. Finally, one compares the fit of the random sample to the fit of the experimentally derived measurements to obtain a probabilistic measure of the accuracy of the fit of the calculated curve. This was Pearson's version of what is now called the χ^2 test (pronounced *kī-squared*); χ is the Greek letter chi. As for so many other discoveries in statistics, the value of the χ^2 test was quickly recognized. Although it is no longer used in quite the way that Pearson preferred, the χ^2 test is still one of the most widely used statistical techniques for testing the reasonableness of a hypothesis.

In addition to Weldon, Pearson collaborated extensively with another scientist, the British eugenicist and anthropologist Francis Galton (1822–1911). As in his work with Weldon, Pearson studied inherited characteristics with Galton. At the time, only one person had any real insight into the nature of heredity, the Austrian scientist and monk Gregor Mendel (1822–84). Mendel had worked hard to discover the mechanism of heredity, and he had published his ideas, but his ideas attracted little attention. In fact, they had to be rediscovered in the early years of the 20th century.

Pearson, as all of his peers did, worked on problems in heredity without any real understanding of the subject. This is both the strength and weakness of statistical methods. In the right hands statistical techniques enable one to discover interesting and sometimes important correlations between different variables. But they are no substitute for theoretical insight. Pearson is best remembered for his insight into statistical methods rather than for what his methods uncovered.

R. A. Fisher

If anyone contributed more to the development of modern statistical thought than Pearson, it was the British statistician and geneticist Ronald Aylmer Fisher (1890–1962). As did Pearson, Fisher had very broad interests. He graduated from Cambridge University in 1912 with a degree in astronomy. This is probably where he first became interested in statistics. Astronomers make many measurements and then use statistical ideas and techniques to interpret these measurements. One of the books that Fisher read at Cambridge was *Theory of Errors* by George Biddel Airy, a prominent British astronomer. He did not work as an astronomer, however. After graduation Fisher worked briefly as a mathematics teacher, but he decided to leave teaching and work as a scientist. He had an opportunity to work for Pearson, who was already prominent in the field, but he turned it down; Fisher and Pearson did not get along. Instead, Fisher was hired as a biologist at the Rothamsted Agricultural Experiment Station.

Rothamsted was founded in 1843 by a fertilizer manufacturer named John Bennet Lawes. In conjunction with the chemist, Joseph Henry Gilbert, Lawes began to make a series of long-term experiments on the effects of fertilizers on crop yield. (Some of these experiments have been running continuously for well over a century.) As the years went by, bacteriologists, chemists, botanists, and others, were added to the staff. All of these scientists running all of these experiments generated a huge collection of data. They were, for the most part, data without a strong theoretical context in which to be placed. In 1919, when Fisher began to work at

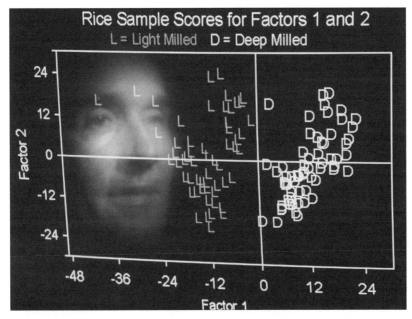

A researcher at the United States Department of Agriculture. Fisher's work on statistics continues to have an important influence on researchers today. (Photo by Keith Weller, ARS/USDA)

Rothamsted, it was the ideal environment for someone with an interest in statistics. (Rothamsted is not as well known today as it was when Fisher began work there. This is, in part, because Rothamsted has merged with other research centers to form the Institute for Arable Crops Research [IACR]. It is, however, still an important research center.)

During his time at Rothamsted, Fisher was highly successful, both as a mathematician and as a scientist. Faced with a surfeit of data, a great variety of questions, and only a weak theoretical context to tie the two together, Fisher began his life's work. He pushed back the frontiers of statistical thinking, and he used his insights to solve certain mathematical problems in genetics. (We will have more to say about this later.) His book *Statistical Methods for Research Workers*, published in 1925, may well be the most successful book on statistics ever written. It was in print for 50 years.

Fisher is often described as one of the founders of modern statistics. As a statistician he made contributions that were very wide ranging. He was the first to investigate the problem of making sound deductions from small collections of measurements. This was a new topic in statistics. Recall that Pearson had used large data sets. In fact, every statistician would prefer to use large data sets—and the larger the better. The reason is that more information, properly understood, cannot lead to less certainty, so one cannot do worse with more. The hope, of course, is that more information will lead to less uncertainty. Despite the desirability of large data sets, the researcher cannot always specify the sample size with which he or she will work. Large data sets are not always available, so having techniques for working with small data sets is valuable.

Drawing sound conclusions from small data sets often involves some fairly subtle reasoning. Consider the problem of finding the mean, or average value, of a character from a small number of measurements. If we calculate the average or mean value of a small set of measurements derived from a few experiments, we can expect the mean to be *unstable:* That is, if we repeat the experiments, make the same measurements, and compute a new mean, we can expect to see large fluctuations in the mean from one set of experiments to the next. For example, if we roll a die three times and calculate the average number points that appear and we repeat the entire procedure several times, we will probably notice significant variation of the mean from one set of trials to the next. The die has not changed, of course; it is just that the sample size is too small to reflect reliably the average we would see if we rolled the die many times. Nevertheless, small sets of measurements are sometimes all that is available. Worse, it is not always possible to expand the set. Under these conditions the researcher must draw the best conclusions possible from the data at hand.

To illustrate how a small data set can contain information about the mean of the much larger (theoretical) set from which it is drawn, let the letter m represent the mean of some larger, unknown set. Suppose that we have only two elements, chosen randomly, from the parent set. We will call those two numbers x_1 and x_2. To be definite, we will suppose that x_1 lies to the right of m. There is a 50/50 chance that x_2 lies to the left of m since, roughly speaking, the average of a set of measurements lies at the center of the set. (Of course, we can say the same thing if x_1 lies to the left of m.) Consequently, there is a 50/50 chance that the mean, or average value, of the parent set lies between any two numbers chosen at random from the set. This is a simple example of the type of deduction that one can make from small data sets, and it is an example that Fisher himself described.

Fisher was also interested in developing more sensitive tests of significance. To appreciate the problem, suppose that we are given two separate sets of measurements. We may imagine both sets of measurements as representing numbers randomly drawn from the same parent set; that, in any case, would be our hypothesis. (If our hypothesis is incorrect then the two samples were drawn from two separate, nonidentical parent sets.) The question, then, is whether or not our hypothesis is reasonable. The difficulty here is that even if the hypothesis is correct, the two samples will almost certainly *not* be identical. Because they were randomly drawn, there will almost certainly be a difference in their average values as well as in their variation about the average. To determine whether the variation that we see between the two sets is significant—in which case our hypothesis is false—or whether the variation is just the result of the randomness of the draw—in which case our hypothesis is true—we need a rational criterion. Fisher was very interested in this problem, especially for the case of small sample sizes, and he made contributions to solving it for various situations.

Another topic in which Fisher had an interest was the problem of experimental design. This is not a topic to which many nonscientists give much thought. Many of us imagine that conceiving the idea for the experiment is the major hurdle to overcome, and that once the experiment is imagined, all that is left is to perform it. In fact, the idea for the experiment is only the first step. An experiment is like a question: The way that one phrases the question has a lot to do with the answers one gets. Furthermore, experiments can be both time-consuming and expensive, so it is important to make the best, most efficient experiments possible since time and money limit the number of experiments that can be performed.

Fisher described some of his ideas about experimental design in a famous, nontechnical article, "Mathematics of a Lady Tasting Tea." This is a very readable introduction to the problems involved in designing productive experiments. In this article Fisher supposes that a woman claims that she can—on the basis of taste alone—determine whether the tea was added to the cup first and then the milk, or whether the milk was added first. The idea is to determine experimentally whether she is correct or incorrect in her assertion. Furthermore, Fisher supposes that he must make his determination after exactly eight "taste tests." (This is a small sample size.)

Initially, Fisher supposes that he will offer the lady four cups in which the milk has been added first—we will call this MT (milk then tea)—and four cups in which the tea was poured first—we will call this TM (tea then milk). Furthermore, he assumes the lady is also aware of this restriction, so she too knows her answers must contain four "milk then tea" (MT) and four "tea then milk" (TM). There are 70 different ways of ordering four MT symbols and four TM symbols. Because there are 70 different possible answers, there is a 1/70 chance that she will simply guess all the right answers. Fisher then compares this design with other possible experimental designs consisting of eight taste tests. He points out that if the lady is simply offered eight cups, where the number of MTs can vary from 0 to 8, and the remaining choices are TMs, then the lady has only a 1/256 chance of guessing the correct answer. This may appear to be a more discriminating test, but if she is offered eight consecutive MTs, then she and the researcher miss their opportunity to test whether she can really distinguish the two tastes. In the search for a more discriminating test, she is deprived of the opportunity to demonstrate her remarkable ability. This kind of situation is characteristic of much experimental design. In particular, there is no one best way of performing an experiment. The design that is eventually chosen always reflects personal preferences as well as research aims. The goal of Fisher's research in this area was to develop rational criteria to evaluate the success of various experimental designs.

Fisher goes on to discuss other techniques for increasing accuracy and for minimizing other factors that may adversely affect the outcome of the experiment. For example, if all cups are prepared simultaneously, and if the woman is presented with all four MTs at the end of the test, then the experimenter has established a correlation between a cooler drink and the "milk then tea" order. This must be avoided because we wish to test her ability to distinguish MT from TM, not warm from cool. (If she happens to prefer one temperature to another, this may well affect her decision about the order in which the liquids that made up the drink were poured.)

Fisher's method of minimizing the impact of these additional factors is not to eliminate them all—this is impossible. Instead, he advocates randomizing every conceivable variable except the one of interest. The hope is that randomizing all factors except the one being studied will cause the effect of all other factors on the outcome of the experiment to diminish and the desired factor to stand out. All of these various complications point to the fact that designing a fair experiment is difficult. Though the tone of this article is lighthearted, "Mathematics of a Lady Tasting Tea" presents a nice insight into the kinds of problems one encounters whenever one designs an experiment.

Fisher used many of these ideas in his study of population genetics. In addition to helping establish the field of modern statistics, Fisher also did a great deal to establish the field of population genetics. His book *Genetical Theory of Natural Selection* is one of the great classics in the field.

Population genetics is a highly mathematical branch of genetics. It centers on the problem of mathematically modeling the distribution of genes within a population and calculating how gene frequencies change over time. Scientists in this discipline are sometimes interested in providing mathematical explanations for how certain complexes of genes arose, or they may be interested in describing how the frequency of certain genes changes in response to changes in the environment. The emphasis in population genetics is less on the individual—whose genes cannot change—than on the species to which the individual belongs. (Genetically speaking, individuals do not change—we die with the

Charles Darwin. Fisher synthesized the ideas of the naturalist Charles Darwin and the geneticist Gregor Mendel and expressed them mathematically. (English Heritage/Topham-HIP/The Image Works)

genes with which we were born; the genetic makeup of species, however, does change over time.)

Fisher was well placed to contribute to this field. Gregor Mendel's work had, by this time, been rediscovered and improved upon. Many advances were being made in understanding the way traits are inherited, and there was a great deal of interest in applying these insights to the field of natural selection. Even early in the 20th century, many scientists had accepted the validity of Charles Darwin's ideas about how species change over time but were unsure of how traits were passed from one generation to the next. Understanding the mechanism of heredity from a statistical point of view was critical if scientists were to understand how, and how fast, changes in the environment affected gene frequencies. Fisher sought answers to these problems through an approach that used probability and statistics.

One type of problem that is important in population genetics is determining how changes in gene frequencies are related to population size. Population size is a critical factor in determining how a species changes from one generation to the next. A species with many individuals is better able to harbor many different combinations of genes; therefore, when the environment changes, it is more likely that there will be some individuals already present who are better adapted to the new environment. This idea is an important part of evolutionary theory. Although large species are generally

better able to adapt to rapid changes, scientists are not always interested in studying the species with the most individuals. Sometimes
the species of most interest are the most rare. Developing methods
for drawing reliable conclusions about small populations becomes
critical for understanding the evolution of rare species. Some of
these methods were developed by Fisher himself.

Fisher remained at the Rothamsted Agricultural Experiment
Station from 1919 to 1933. He then moved on to the University
of London and later Cambridge University. In his last years of life,
Fisher moved to Australia, where he continued his work. It is not
often that one encounters someone who is so successful in two distinct fields. Fisher's contributions to statistics and the genetic basis
of evolutionary change are especially noteworthy because he made
them at a time when most scientists were drawn to ever-increasing
specialization.

Fisher and Pearson did much to establish the foundation of
modern statistics. To be sure, the types of problems on which they
worked are elementary by modern standards. Mathematics has
advanced considerably in the intervening years, and so has computational technology. (By the time the first computer was constructed Pearson had died, and Fisher was in his 50s.)
Nevertheless, many of the ideas and techniques that they developed are still used regularly today in fields as diverse as the insurance industry and biological research.

11

THE THEORY OF SAMPLING

There is a tendency, especially in a history of mathematics, to convey the impression that mathematics was invented by mathematicians. This is not entirely true. If we look at the biographies of prominent mathematicians in this series we see that Girolamo Cardano was a physician. Galileo Galilei, Edmund Halley, and Isaac Newton were physicists. Rene Descartes, Blaise Pascal, and Gottfried Leibniz, though very important to the history of mathematics, were more philosophers than mathematicians. Pierre Fermat and François Viète were lawyers. Marin Mersenne, Thomas Bayes, John Wallis, and Bernhard Bolzano were members of the clergy. John Graunt was a businessman. Karl Pearson was a social activist, and R. A. Fisher was a geneticist. Of course, all of them contributed to the development of mathematics, but whether they were specialists in mathematics—that is, mathematicians as we now understand the term—or whether their principal interests and energies were directed elsewhere is not always so clear.

One very important branch of statistics, sampling theory, was developed largely by people for whom mathematics was almost an afterthought. Their discoveries are vital to a variety of applications, including the ways in which societies make decisions about the distribution of resources, television networks decide which shows to keep and which to cancel, seed companies improve their seed lines, and national economies are managed. The theory of sampling as a separate discipline within statistics began, however, with problems related to manufacturing, and so, historically speaking, it is a fairly recent invention.

The Problem

To appreciate how the theory of sampling arose, it helps to know a little about the problems it was designed to address. For most of human history every object was hand-crafted. Each complicated object was made by a specialist. There have been exceptions. The Egyptians, who constructed their largest, best-known pyramids over the course of just a few centuries, must have cut, transported, and piled the millions of large stone blocks required to make these monuments in a way that used at least a few assembly-line techniques. They also apparently standardized the construction of bows and arrows so that the weapons of one soldier were interchangeable with those of another. For the most part, however, societies were neither large enough nor organized enough to require the sort of mass production technology that dominates modern life.

The situation began to change during the Industrial Revolution. One of the first critical suggestions was offered by Eli Whitney, who is celebrated in American schools as the inventor of the cotton gin long after the purpose of a cotton gin has been forgotten. (A cotton gin is a device for separating cottonseeds from cotton fiber. Its invention made cotton the principal cash crop in the southern United States for many years.) Whitney suggested that guns—flintlock guns, to be specific—be manufactured in such a way that the parts from different rifles could be interchanged with one another. In 1798 the federal government awarded him a contract to produce 10,000 muskets using his Uniformity System, an early version of mass production.

This approach was in stark contrast to traditional methods. Previously, each part of a gun was created to fit the other parts of a particular gun. Of course, all guns of a certain type had characteristics in common. They were of roughly the same dimensions, and they worked on the same basic physical principles. But it was usually not possible to use one gun as a source of spare parts for another, even when both guns were created at roughly the same time by the same craftsperson. There was too much variation in the product. Whitney's new method of manufacturing guns was meant to overcome this shortcoming, but it also pointed to a new method of manufacturing other objects as well.

This new manufacturing method required standardization of design and materials. It also required that the manufacturing process be sufficiently controlled that an object made at the beginning of the week would be "identical enough" to the object made at the end of the week to make them interchangeable. This change in the concept of manufacturing has changed the world. We are still grappling with its implications for labor and for our standard of living.

Throughout the 19th century, industrial engineers on both sides of the Atlantic worked to implement the new ideas. "Simple" objects such as nuts and bolts, textiles, and pulleys began to be manufactured according to standard designs using methods that greatly increased the quantity of finished goods, where the quantity is measured both in numbers and in the number of units per person engaged in the manufacturing process. Nor was this increase in production due solely to what was happening on the factory floor. Much of the culture and technology of the time was aimed at facilitating manufacturing operations. Steamships were plying the world's oceans transporting raw materials to the manufacturing sites and finished goods from the manufacturing sites to consumers around the world. A great deal of money was changing hands, and this served to accelerate progress further.

Late in the 19th century the American engineer and inventor Frederick Winslow Taylor (1856–1915) began to search for more efficient production processes. Taylor, who had a degree in engineering from Stevens Institute of Technology, Hoboken, New Jersey, was interested in improving the human processes by which goods were manufactured. He studied the physical motions of workers involved in a manufacturing process and sought to streamline them. He called his ideas "scientific management." Taylor was quite successful. As a consequence of his work, productivity—the amount of goods produced per worker—soared again.

Many of these "hard" and "soft" technologies meshed together in the mind and factories of the American industrialist Henry Ford (1863–1947). Ford was engaged in producing what was, and arguably still is, the most technologically sophisticated consumer

Interior of the tool and die building at the Ford River Rouge plant, Dearborn, Michigan, 1941. Mass production of increasingly sophisticated goods demanded new statistical tools to establish and maintain control over industrial processes. (Office of War Information, Library of Congress, Prints and Photographs Division)

item ever made, the automobile. To accomplish this he joined techniques of mass production and scientific management to produce huge numbers of cars at a cost that many people could afford. This was a tremendous technical accomplishment because it involved the coordination of large numbers of workers, the acquisition of huge numbers of parts, and the design of an industrial process such that the quality of the final product was controlled from day to day and week to week.

Ford's manufacturing technologies were quickly emulated and improved upon in many places around the world. Large, complex manufacturing concerns were producing ever-increasing amounts of consumer goods. As the complexity of the manufactured goods increased, controlling the quality of the items produced became increasingly difficult. How could one maintain control of the various processes and materials involved so that the quality of the finished product was uniform? How, in effect, could the right hand know what the left hand was doing?

Walter Shewhart and Statistical Quality Control

Walter Shewhart (1891–1967) was the first to present a complete and coherent approach to the issue of quality control. He was an American who was interested in science and engineering. He attended the University of Illinois and earned a Ph.D. in physics from the University of California at Berkeley in 1917. It is, however, hard to categorize Walter Shewhart. Because his goal was to secure "economic control" over manufacturing processes, he had to have a thorough knowledge of statistics, economics, and engineering. He was the first person to create a comprehensive and unified treatment of statistical quality control. His accomplishment was quickly recognized by some and ignored by others. Many encyclopedias and other reference books fail even to mention Shewhart. On the other hand, various quality control organizations often credit him with inventing the concept of statistical quality control. There are an increasing number of tributes to him on the Internet, some of which are more like shrines than biographies. His friend Edward Deming remarked as late as 1990 that it would be 50 years before Shewhart's contributions would be widely understood and appreciated. There is little doubt that his contributions were not fully appreciated for the first half-century after Shewhart published his ideas.

After earning his Ph.D., Shewhart worked briefly as an academic, but he soon left for a job at Western Electric Company. Six years later he found a position at Bell Telephone Laboratories, one of the premier scientific research establishments of the 20th century. He received numerous awards and honors as his ideas became better known. He sometimes lectured at academic institutions, including the University of London, Stevens Institute of Technology, Rutgers, and a variety of other institutions, academic and industrial, but he remained employed at Bell Labs until his retirement in 1956.

Shewhart wrote that early (19th-century) attempts at mass production had tried to eliminate variability; a more effective goal, he said, was to *control* variability. The idea apparently occurred to him while he was working at Western Electric. Western Electric manufactured telephony equipment, such as amplifiers, for Bell Telephone. Telephony equipment was often buried underground at this time, so for economic reasons equipment failure rates had to be reduced to an absolute minimum. This was a major goal at Western Electric. The company had made great efforts to improve the quality of their product, and for a while they made progress in this regard. Eventually, however, despite the fact that the company was still spending a great deal of money and effort on improving the quality of its manufactured goods, progress in quality control began to slow. Undeterred, the manufacturer continued to search for mistakes in production. When engineers found them, they adjusted the manufacturing process to eliminate the cause of the problem. They did this again and again, and over time Western Electric engineers noticed a remarkable fact about their efforts at quality control: As they adjusted control of the manufacturing process in response to the defects their testing had detected, the quality of the manufactured product actually *decreased*.

The quality control problems at Western Electric did not result from lack of effort; the problem was lack of stability. A stable manufacturing process is vital for a quality product. Western Electric management, however, did not need only to find a method for detecting instabilities in quality; more importantly, they also needed to identify the causes of variation. Causes are critical. Shewhart hypothesized that variation has two causes: random or *chance cause* variation, which is also called common cause variation, and *assignable cause* variation, which is also called special cause variation. Assignable cause variation is variation that can be identified with a specific cause—machinery that is out of alignment, variation in the quality of materials, sloppy work habits, poor supervision— anything that can be associated with a cause. Assignable cause variation can be eliminated, and the goal of the manufacturer was, according to Shewhart, to eliminate assignable cause variation. After assignable cause variation was eliminated, only the random or chance cause variation was left. There is *always* some random variation associated with any process. The goal for chance variation was, therefore, not to eliminate it, because that would be impossible. The best that one could hope for was to control it. Any manufacturing concern that could eliminate assignable cause variation and control random variation—and do this in an economical way—could exert economic control over the manufacturing process.

This is a very delicate problem. The goal is not simply to identify faulty pieces and reject them before they reach the end of the assembly line, because every rejected piece costs the company money. Simply identifying the rejects does not place the process under *economic* control. Instead, the manufacturer must manufacture as few faulty units as possible. Second, and just as important, the manufacturer must be able to recognize when this level of efficient manufacturing has been attained. Although this may seem simple to do, it usually is not.

Everything in Shewhart's analysis of production control is subject to the sometimes-conflicting constraints of economy and quality. Consider, for example, the problem of producing air bags for automobiles. The bags must deploy under the right circumstances, and they must not deploy when they are not needed. Control of the manufacture and installation of air bags is vital. The best way to test the air bags is to install them and crash the cars in which they are installed. Unfortunately, this type of testing can be very expensive. Worse, it destroys the air bag being tested as well as the car in which it is installed. The conflict between economy and reliability testing is, in this case, obvious. Once the manufacturer has identified when it has begun to make a product that is largely free of defect, it must be able to maintain the stability of the process. Finally, the manufacturer must also be able to place the manufacturing processes under economic control.

When a manufacturer identifies faulty product, how can it determine whether the defects are related to assignable cause variation or random, chance cause variation? One of Shewhart's most useful ideas in this regard was the control chart. This idea has been refined quite a bit, so that today there is not one control chart—sometimes called a Shewhart chart—but many. Despite this diversity, there are three basic components of every control chart. The first component is a horizontal line across the center of the chart that represents the average or mean value of the property being measured. Second, there are a line above and a line below the centerline. These lines define the upper and lower bounds for acceptable variation. The third basic component is a record of the data collected. The data are plotted over time. See the accompanying diagram to see how a control chart might look.

In manufacturing processes in which economic control is exerted, the points on the control chart should be randomly distributed. There should be, for example, no "general trends" in the data: That is, there should be no runs of continually increasing or decreasing sets of points. Similarly, there should be no long runs of points that all fall on one side or the other of the centerline. Nor should there be a long run in which the points

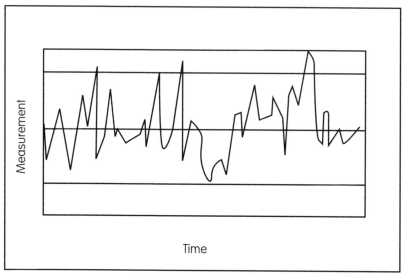

Model Shewhart control chart. Shewhart sought ways for manufacturers to place their processes under economic control.

regularly alternate from one side of the line to the other. These patterns are good indicators of nonrandom, or assignable cause, variation in the product quality—the type of variation that must be removed from the manufacturing process. This is one of the great advantages of using a control chart: It enables the user to identify assignable cause variation given only the results of tests on the product.

There is one more characteristic of a process that is under economic control. It is, perhaps, the easiest to notice: All of the data on the chart should remain between the upper and lower bounds of variation. The control chart makes visible the problem of economic control of a manufacturing process. Once it has been established that the process is under economic control, the next step is to reduce the random variations in product quality.

Sometimes even more constraints are placed on the chart than those mentioned here. Those constraints can make use of the charts problematic, because they disturb the balance between costs and quality. If too many constraints are placed on what acceptable data should look like, then the control chart will sometimes detect fictitious control problems. Keep in mind that there is always some variation in the manufacturing process. It cannot be prevented. If the control chart yields a "false positive" finding on the issue of assignable cause variation, for example, then management has no alternative but to begin to reevaluate the manufacturing process and to make whatever changes are needed to eliminate the variation. If, however, the variation is fictitious, then doing so is, at minimum, a waste of time. Recall, however, what Western Electric engineers discovered so long ago: Adjusting control of the manufacturing process has, in the short term, the effect of decreasing quality rather than increasing it. Consequently, false positive findings need to be prevented.

The advantage of using a simpler control chart is that it is less likely to detect false positive findings, in part because simpler control charts are less likely to detect anything at all. On the other hand, an insensitive chart may well miss actual assignable cause variation, the identification of which is one of the advantages of the chart.

As with most ideas in statistics, the successful use of this concept involves subjective judgments as well as science. Nevertheless, the control chart quickly found its way into many manufacturing concerns. It is a straightforward expression of a deep insight into the idea of quality control. Despite its simplicity it enables large and complex organizations to separate random from assignable cause variation and to monitor the former while eliminating the latter. It was a very big breakthrough.

Nonetheless, it would be inaccurate simply to identify Walter Shewhart with his control chart, as is sometimes done. His control chart is a nice illustration of his ideas, but his ideas go deeper than is immediately apparent. He combined statistics with economics and engineering to create the branch of knowledge now known as quality control or statistical quality control. Shewhart's ideas quantify what is meant by variation in manufactured goods, and by variation in general. As manufacturing tolerances become ever more exact, Shewhart's ideas have remained valid. They have made the existence of large, complex, well-regulated manufacturing concerns possible.

William Edwards Deming

Shewhart's control charts can be applied to practically any process managed by any organization, public or private, provided the product created by the organization can be unambiguously represented with numerical measurements. Of course, any technique has its limitations. It is worth remembering that there are some services, usually managed by governmental or not-for-profit organizations, in which the "product" is not so easy to measure. The care of the severely disabled, for example, is difficult to quantify because the product, which is the standard of care, does not easily translate into straightforward numerical measurements. Every long-term, profit-making enterprise, however, is amenable to analysis using Shewhart's control charts because every profit-making organization has a bottom line: A company's profit–loss statement is, in the end, a numerical description of the success or failure of the company in terms that are clearly understood by all interested parties. It may seem, then, that Shewhart, in a very general way, solved the problem of quality control. He almost did. He gives a clear statement of the control chart. The statistical procedures necessary to analyze the data on the chart can be obtained from the work of Pearson and Fisher. There was, however, still one missing piece: a comprehensive theory for obtaining representative samples. This was one of the contributions of the American statistician and philosopher William Edwards Deming (1900–93).

Deming was born into a poor family. His family moved several times during his youth as his father searched for employment. It was not an easy life, but the son was ambitious. He obtained a bachelor's degree from the University of Wyoming, where he majored in electrical engineering, and a Ph.D. in mathematical physics from Yale University in New Haven, Connecticut. (He was especially interested in the problem of Brownian motion. See Chapter 5 for a discussion of this random phenomenon.)

Deming worked his way through school. This might have seemed like a hardship at the time, but in retrospect it turned out to be the best thing that could have happened to him. For two summers Deming worked at the Western Electric Company in Chicago. Walter Shewhart was also working at Western Electric at the time, and the two met. Western Electric had been struggling with quality control problems, and Shewhart had begun to think about the ideas of chance variation and assignable cause variation. Deming had arrived at exactly the right time, and, apparently, he recognized it. Shewhart became Deming's mentor, and Shewhart's discoveries formed the basis of much of Deming's later research. He never forgot his debt to Shewhart. Never an arrogant man, Deming was still quoting and praising Shewhart's 1931 masterpiece, *Economic Control of Manufactured Product*, in his public speeches more than 50 years after the publication of Shewhart's book.

Despite his summer jobs at Western Electric and his great admiration for Shewhart's ideas, Deming never again worked in a manufacturing environment. His first job after completing his education was with the U.S. Department of Agriculture. The department provided a rich source of problems for someone interested in statistics, and Deming remained there for more than 10 years. In 1939 he moved to the U.S. Census Bureau. His work at the Census Bureau changed his life and, as we will soon see, may have changed the world. One of Deming's duties at the Census Bureau was to provide guidance on problems associated with sampling.

Sampling theory is a branch of statistics. It is concerned with the problem of obtaining a sample, or subset, from a larger set in such a way that one can make accurate deductions about the makeup of the larger set by analyzing the sample. This is a central aspect of constructing a good control chart. The product must be sampled in such a way that the objects tested are a good representation of the entire set of objects. Designing a method that ensures accurate sampling is not an easy problem. Mathematically, the difficulty arises because one does not know the properties of the larger set. (If one did, then it would not be necessary to sample it.) The sampling is further constrained by costs and time factors. This is certainly true of the work at the U.S. Department of Agriculture. Deming's work in the 1940 census was highly praised, and it resulted in *Some Theory of Sampling*, published in 1950 and still in print 50 years later, and an invitation to Japan to assist in the first postwar Japanese census.

Deming's *Some Theory of Sampling*, despite the inclusion of the word *some*, is a very hefty book about many aspects of sampling theory. The theory is described from the point of view of practical applications. In his book Deming gives careful attention to the concepts that make successful sampling possible. One of these is the importance of carefully defining the *universe*, or parent set, from which the sample is drawn. Sometimes the universe is given as part of the problem. In an industrial operation the universe is the set of all objects produced during a production run. Other times—for example, when trying to sample the set of all consumers considering buying an automobile in the next year—the universe is more difficult to specify with accuracy sufficient to be useful. Without a clear and unambiguous definition of what a universe is, it is not possible to obtain a representative sample of it.

Once the universe is defined, the next problem is to develop techniques for obtaining a representative sample from it. No problem could be more important for obtaining an accurate statistical description of the parent set. This problem is, for many applications, far from solved even now. For example, in certain regions of the country predictions of the outcome of the 2000 presidential elections and the 2002 midterm elections were inaccurate. Researchers defined the universe of likely voters and sampled the voting preferences of this group for statistical analysis, but in certain areas the actual voting patterns were different from those predicted. Even now, it is not entirely clear why the analyses were inaccurate. One possible conclusion is that the universe of all voters in each of these elections was different from the universe the researchers defined. Another possible conclusion is that the statisticians correctly identified the set of likely voters but failed to obtain a representative sample.

In his book Deming carefully considers the problem of sampling error, but statistics is an applied field of knowledge; the problems associated with sampling theory are practical as well as theoretical. To be sure Deming covers the mathematical ideas associated with the subject, but he is equally interested in the problems associated with obtaining a sample economically and quickly. His conception of the field of sampling theory is characteristically broad.

Deming's presentation is heavily influenced by his time at the Census Bureau and the Department of Agriculture. For example, sometimes the Department of Agriculture carried out preliminary surveys whose results depended on sampling theory before making an exhaustive survey, that is, a survey that polled every single element in the parent set. This enabled the statistician to compare predictions based on samples with the results from the more exhaustive surveys. These surveys are analyzed in some detail. When the results were good, the surveys could be used to guide future statistical work; when the results were bad, the surveys served as examples of approaches to avoid. Deming's extremely pragmatic overall approach to the theory of sampling was threefold: (1) to specify the reliability of the survey to be carried out— that is, the precision desired; (2) to design the survey so that it will achieve the sought-after precision at the least possible cost; and (3) to appraise the results.

Deming's approach is remarkable for both its clarity and its emphasis on economics. Money and time are issues in any scientific undertaking. This has always been true, but it is not until Shewhart and Deming that we see scientists so explicitly linking the search for knowledge to economic considerations. We are accustomed to these ideas now. They are contained in any cost/benefit analysis undertaken by any institution, public or private. Shewhart and Deming, however, were pioneers in formulating these concepts in a way that was explicit enough for planners and policy makers to manipulate them to get the greatest benefit for the least cost, whether that cost is measured in time, money, or both.

At about the same time that *Some Theory of Sampling* was published, Deming was invited to Japan to assist in the national census there. On his second census-related visit to that country, Deming was invited to give several lectures to Japanese industrialists, engineers, and plant managers on methods for achievement of quality in industrial production. In one particularly important meeting a number of Japan's leading industrialists attended one of his lectures. Deming's ideas on the importance of Shewhart's charts and the statistical theories involved in their implementation made a huge impression, and these ideas were implemented in Japanese industry long before they found a home in American or European industry. At that time Japanese industry was being rebuilt after the devastation of the Second World War. Japanese industrialists took Deming's advice to heart because it offered the possibility of a rational and superior method for controlling variation in the quality of manufactured goods.

It is worth noting that within a generation after their destruction, Japanese industries were famous throughout the world for the very high level of economic control that they had learned to exert over the manufacturing process. In certain important industries, this advantage is maintained to the present day. For the remainder of his life, Deming always received a warm and enthusiastic audience among Japanese academics and industrialists, but for a long time his ideas were not recognized in the West.

By the 1970s North American and European industries that were in direct competition with Japanese industries had begun to see severely eroded profits and market share. These industries were often selling less even as demand for their type of product increased; they were being crowded out by more successful Japanese firms. By the 1980s, Shewhart's and Deming's statistical approach to quality control was drawing a great deal of interest from Western industrialists. Shewhart had died decades earlier, and Deming was already in his 80s, but he was determined to spread the word. He developed a famous four-day seminar that he delivered throughout the world at academic, government, and business institutions. In it he described the difference between random variation and assignable cause variation, explained the use of Shewhart's control charts, and gave a very inclusive definition of statistical quality control management. This time people in the West were willing to listen and sometimes to follow his advice.

As his health failed Deming also began to present the lecture "A System of Profound Knowledge." His lecture incorporated the theory of variation, psychology, a theory of knowledge, and system theory, which involves insights into what organizations are, how they make decisions, and how they work the way they do.

William Edwards Deming and Walter Shewhart found a way to use statistical thinking to create a new way of understanding processes. Their insights began with manufacturing processes, but their influence has spread beyond that realm. Because the value of their insights was recognized in Japan, Deming and Shewhart contributed substantially to the postwar growth of Japanese industry. The success of Japanese industry initially caused a great deal of economic hardship in North American and European industries as companies in those regions struggled to identify their problems and correct them. Over time, however, the ideas of Shewhart and Deming have become central to the control of quality and the efficiency of production in industries throughout the world. They have changed the lives of many people around the globe.

11

THREE APPLICATIONS
OF STATISTICS

Statistics is an applied science that developed in response to specific problems. Unlike in many branches of science, in which the majority of expert practitioners are to be found in academia, many of the most knowledgeable statisticians are to be found in positions of government and industry. This is just a reflection of the importance of the discipline to the formation of rational policy. To obtain some feeling for the scope and history of the subject, it is helpful to see how statistics has been applied to solve problems in health, government, and marketing.

The Birth of Epidemiology

Epidemiology is a branch of medicine that is concerned with the incidence, spread, and control of disease. It depends in a critical way on the use of statistics to give insight into how disease is transmitted from individual to individual and place to place. Epidemiology is a fairly new branch of medicine. Its first great breakthrough was in the fight against the disease cholera. This famous application of statistics offers insight into the strengths and weaknesses of statistical methods.

Cholera is an ancient disease. It is caused by a bacterium, called *Vibrio cholerae*, that for most of the history of humankind remained within the borders of the Indian subcontinent. Cholera is a dramatic disease. Its onset is sudden and violent, and the disease is sometimes fatal. The death rate of cholera varies widely from area to area and outbreak to outbreak. It can be much higher or much lower than 50 percent.

Cholera is a disease of dehydration, and its progression has often been described as occurring in three stages. The onset of the disease is marked by violent vomiting and severe diarrhea. In a brief period the infected person can lose as much as 10 percent of his or her body weight. The first stage concludes with what early-19th-century physicians called "rice water diarrhea," which consists of a clear liquid containing what we now know are fragments of the lining of the patient's damaged intestine. During the second stage, which these physicians called the "collapse stage," vomiting and diarrhea cease, the body temperature drops, the pulse becomes very weak, lips and fingernails turn blue, and the blood is very thick and dark—almost black. It is during the second stage that the majority of deaths occur. The third or "recovery stage" is marked by fever.

Outbreaks can be devastating. In 1781 in the Indian city of Haridwar, which is located on the banks of the Ganges River, there was an outbreak of cholera in which 20,000 individuals died in eight days. Although this was a particularly severe outbreak, it was typical in the sense that it was confined to the Indian subcontinent. In the year 1817 the situation changed.

The more recent history of cholera is described in terms of pandemics, outbreaks of the disease that occur over very large geographical areas. The first pandemic began in 1817. In that year cholera spread across the Indian subcontinent. By 1819 cholera could be found in what is now Sri Lanka. By 1820 East Africans were dying of cholera for the first time. By 1821 cholera was present on the Arabian Peninsula, and by 1822 Japan and China were suffering from cholera. In 1823 cholera began to make its way into Russia, with an outbreak in the city of Astrakhan in which there was a total of 392 reported cases and 205 fatalities. The disease then disappeared everywhere except the Indian subcontinent, where it is endemic. During this time not a single effective strategy for combating cholera was developed.

The second pandemic lasted from 1826 until 1837. Many historians believe that the second pandemic began at Haridwar at one

Nineteenth-century cartoon satirizing the sometimes-ridiculous measures taken by people to protect themselves from cholera. Before Snow's work, however, no one understood how cholera was transmitted. The efforts of many physicians of the time were, in retrospect, no more effective than the efforts of this figure. (Library of Congress, Prints and Photographs Division)

of the great religious festivals that occur there. It is thought that religious pilgrims from Bengal introduced the disease to the festival at Haridwar, and from Haridwar it was dispersed throughout India. The disease followed rivers and trade routes. Again it appeared in the Russian city of Astrakhan. Within 24 hours of the first reported cases there were 200 deaths. Among the first to die were the civil governor and the chief of police. Panic ensued. The social order collapsed as much of the population fled into the surrounding countryside, taking the disease with them. Between July 4 and August 27, 1830, there were 3,633 cases of cholera recorded at Astrakhan. About 90 percent of those who became ill at Astrakhan died.

Cholera made its way northward along the Volga River and then spread out. As it did so it became less deadly. By the time it reached Moscow the death rate was 50 percent of those who became infected. The Moscow outbreak attracted a great deal of attention throughout Europe, which had never had a cholera outbreak before. From Russia, cholera moved for the first time into the Baltic states and Poland. From Poland it moved from country to country as far west as France and Great Britain. It crossed the Atlantic Ocean, and for the first time cholera appeared in Canada and New York and then moved south and west. By the time the outbreak was over in 1837, cholera had become a worldwide epidemic.

Cholera inspired a great deal of fear, but statistical techniques were not applied to the study of cholera until the third pandemic (1846–63). Medically speaking, the second pandemic was noteworthy because a successful treatment for cholera was first developed at this time. Unfortunately, physicians were not conducting carefully designed statistical experiments to compare various methods of treatment. Some applied leeches for the treatment of cholera, some administered laxatives, and one British physician, Thomas Latta, used a saline fluid injection to rehydrate patients who were on the verge of death. Since some people always get better, even with leeches, and since there were no objective statistical criteria to compare the different approaches, Latta's innovation was ignored by the medical establishment of his time.

Although a treatment of cholera is important, it turns out that the prevention of cholera is an easier, more economical way of protecting the public health. This was discovered during the third pandemic well before the germ theory of disease was discovered, and it shows in a very dramatic way both the power and the weakness of statistical methods. The scientist who made this discovery was the British physician John Snow (1813–58).

John Snow was the son of a farmer. He attended school until he was 14 and then was apprenticed to a surgeon. During the second pandemic he served as an assistant. He was ambitious, and over the course of many years he worked his way up into the medical establishment. He graduated from the University of London in 1844, and in 1850 he was admitted to the Royal College of Physicians. Snow thought long and hard about the problem of cholera. He was especially interested in the process through which cholera is transmitted. He learned as much as he could about outbreaks that occurred during the second pandemic, and he studied previous patterns of transmission. He was as thorough as one man working alone could be.

The geographic patterns of transmission of cholera were quite complicated. Sometimes it struck one area and entirely missed neighboring population centers. Its transmission as a function of time was also very complicated. For example, in 1848, during the third pandemic, cholera caused 1,908 deaths in England and Wales. In 1849 there were 53,293 deaths due to cholera, but during the next two years no deaths in either England or Wales were attributed to cholera. Cholera returned in 1853 and 1854, when there were, respectively, 4,419 and 20,097 deaths attributed to the disease. Although the third pandemic continued until 1863 there were no further deaths in England or Wales during the pandemic.

John Snow began his studies of the mechanism of cholera transmission under a serious handicap: Although no one understood how cholera was transmitted, a number of prominent individuals championed incorrect ideas. When Snow proposed his idea that cholera is transmitted through drinking water, there was no shortage of individuals to tell him that he was wrong. To be sure, Snow had collected and analyzed as much information on past outbreaks as he could. In 1849 he even published "a slender pamphlet" that described his theory, but it did not sway many minds. Instead, his idea that cholera, or at least the cause of cholera, was in the water was greeted with a great deal of skepticism. He did not become discouraged, but he realized that to convince others, and to save lives, he would need to acquire better data. Snow's big break-through occurred during a cholera outbreak in the Broadstreet area of London, beginning on August 31, 1853.

It was an especially virulent outbreak. During the first three days 127 people living in this compact neighborhood died. Within a week of the outbreak of the disease the majority of the survivors had locked their homes and businesses and fled. By September 10 there were 500 fatalities.

Meanwhile, Snow had already begun his investigation. The compactness of the outbreak gave Snow some hope that he could link the fatalities to a single polluted water source. He diligently began to interview families of victims on the first day of the epidemic. He discovered that most of the families of the deceased got their water from the pump on Broadstreet, but several deaths were more difficult to explain. An elderly woman living some distance from Broadstreet also died of cholera. She was an isolated case in her neighborhood with no apparent connection to the outbreak on Broadstreet. Snow discovered, however, that she had once lived in the Broadstreet area and liked the taste of the water enough to have a bottle of it taken to her daily. All of his detective work, his statistical correlations, and his theories failed to convince the authorities, however. By the time they reluctantly agreed to remove the handle from the Broadstreet pump, the epidemic was already substantially over.

Snow received additional help when a minister in the area, the Reverend Henry Whitehead (1825–96), decided to investigate the outbreak himself. Initially, Whitehead did not accept Snow's theory. Determined to prove him wrong Whitehead interviewed as many people as he could—occasionally he visited them several times—until he had found each victim's name and age and had determined whether the individual drank water from the Broadstreet well, the hour the illness began, and what the sanitary conditions were. It was an enormous amount of information. More than 600 people had died over the course of the outbreak. The results of Snow's and Whitehead's efforts were a collection of tables, a complete analysis by Snow, and a map relating the locations of the cases to the position of the well. They even discovered the initial case: A five-month-old girl had contracted cholera shortly before the onset of the outbreak. Her family's cesspool was located three feet from the pump.

Snow died before the germ theory of disease was accepted. He died, in fact, before his theory that cholera is transmitted through drinking water was widely accepted. The statistics did not identify what cholera was. Snow's theory was not an explanation in a scientific sense. It was a correlation between water and disease. Correlations, not explanations, however, are what statistical research excels at revealing.

In 1866, during the fourth pandemic there was a cholera outbreak in London. A government official and statistician named William Farr, who was familiar with Snow's theory on the mechanism of transmission of cholera, examined the water supply to the area. He traced the water consumed by those individuals infected with cholera to some ponds used by a local water supply company. Farr found these ponds polluted with sewage. He used his influence within the government to prevent the company from distributing water from the pond. The epidemic quickly died out. This was the first time that statistics enabled a government to halt an epidemic.

The U.S. Census

The actual Enumeration shall be made within three Years after the first Meeting of the Congress of the United States, and within every subsequent Term of ten Years, in such Manner as they shall by Law Direct. (The United States Constitution, article 1, section 2)

The United States Census is one of the great statistical studies carried out anywhere. Originally, the census began with a simple goal. After the American War of Independence the American states formed a loose association called a confederation. The structure of the new country was described in a document called the Articles of Confederation, and from 1781 until 1789 the new nation struggled to function with the system of government defined by the articles. The experiment was a failure. Under the Articles of Confederation the central government was too weak to govern.

In 1787 delegates from the various states met to correct the problem. One of the more contentious issues they faced was the problem of representation. States with large populations wanted representatives apportioned on the basis of population. Not surprisingly, states with small populations expressed concern that they would be ignored under this system of government. The compromise to which the delegates finally agreed included one legislative body, the House of Representatives, in which each state would be represented by a number of delegates roughly proportional to its population, and a second legislative body, the Senate, in which each state would be represented by two delegates regardless of population. This compromise required fairly detailed information about the size and distribution of the U.S. population, so the men who drafted the Constitution included a section mandating that a

It is difficult to appreciate the scale of contemporary United States Census Bureau efforts. (Courtesy of the United States Census Bureau)

census, or count, of the U.S. population be made every 10 years, beginning in 1790. The U.S. government has fulfilled this constitutional requirement every 10 years since 1790 during years of war and peace, prosperity and hardship.

The first census, the 1790 census, set the pattern for succeeding surveys. The census was taken very seriously. Census administrators divided the nation into districts and arranged for a count to be made in each district. All reports were made part of the public record. Hefty fines were established for any census worker involved in filing a false report, and any citizen involved in supplying false information to a census worker faced a substantial fine as well. Information was to be taken from each person at his or her place of residence. Any individual without a fixed place of residence was counted at the location he or she occupied on the first Monday of August. The final tally was 3,929,214.

The United States has always been geographically large, culturally diverse, and economically complex. Even in the 1790s, many thought that a simple head count did not provide enough information about the nation to help establish policies. Policy makers wanted to know more, and a few questions about the characteristics of the U.S. population were added to the survey. Most of this additional information was not required to decide how to apportion representatives; it was extra information that was designed to shed more light on who lived in the United States. For the 1790 census workers counted (1) the number of free white males older than 15 years of age, (2) the number of free white females, (3) the number of slaves, (4) the number of free white males younger than age 16, (5) the number of all other free persons, and (6) the names of heads of households. This was just the beginning. Every 10 years the demand for information about the U.S. population increased. Remarkably, this information was all collected by a small staff and the legions of census takers who were briefly hired every 10 years to collect the information.

The old volumes make for interesting and sometimes uncomfortable reading. Information about race uses the language of slavery decades after slavery was eradicated as an institution. Nor was race the only equal-rights issue. There was an enumeration of "imbeciles" and the "insane." These labels were sometimes used for purposes of repression. Large percentages of racial minorities in some districts were routinely described as insane. To their credit, census officials acknowledged at the time that these data were inaccurate and revealed more about the census taker than the population that was surveyed. Despite their flaws the early censuses reveal a great deal about the characteristics of a rapidly growing nation.

As the nation became larger and more complex, the task of the statisticians became increasingly difficult. This difficulty was not due only to the fact that the nation was getting larger. There was also an ever-increasing demand for raw data and statistical analyses of the state of the nation. Policy makers wanted information about the manufacturing sector, the agricultural sector of the economy, and demographics of the U.S. population so that they could better formulate policy.

Even in the late 19th century and early 20th century the amount of data collected was enormous. The United States Census was one of the first institutions to make systematic use of data processing equipment, the most famous of which is the Hollerith tabulator. For the census of 1890, the superintendent of the Census contracted for six Hollerith tabulators. Individual information, such as age, sex, color, and place of birth, was punched into a card. The cards were then run through the tabulator, which read the card and recorded the information. Data processing equipment increased accuracy and efficiency. It also enabled statisticians to identify correlations between various characteristics of the population more easily. With new technology and better statistical techniques the amount of *information* that could be gleaned from the raw data collected by the census takers continued to increase. Nevertheless, it barely kept pace with the amount of raw data to be analyzed. For more than 100 years all of this was accomplished by a Census staff that was largely organized for a particular census and then disbanded once the report was written.

By 1900 the job of constructing a statistical description of the United States had turned into an enormous task, and a permanent Census Office was established in 1902. In 1903 its name was changed to the Census Bureau. The demand for information

The Hollerith tabulator. Designed for the United States Census, it was a breakthrough in electric data processing equipment. (Courtesy of IBM Corporate Archives)

required that the bureau work continually, taking surveys, analyzing data, and publishing the findings.

By the 1920 census the volume of data that was being collected and analyzed was impressive even by today's standards. It is particularly impressive when one remembers that the computer had not yet been invented. The 1920 census involved a workforce of 90,000 and questionnaires for 107.5 million people, 6.5 million farms, 450,000 manufacturers, and 22,000 mining and quarrying companies. There were 300 million punched cards. The report entailed the calculation of 500,000 "percentages, averages and other rates," and the publication of 12 large, or quarto, volumes of 1,000 pages apiece. The Information Age began earlier than many of us realize.

The decennial census is now only one part of the mission of the U.S. Census Bureau, but it is still an important one. Each question that appears on the decennial census, the one census that is mandated by the Constitution, is required as a matter of law. This information is used to manage and evaluate federal programs and to draw state and federal legislative districts. Some of the information is used to monitor or satisfy legal requirements that have been imposed by U.S. court decisions. It is also routinely used for planning and decision-making purposes by many companies.

Because of the size and importance of the census, the bureau goes to great lengths to collect and analyze information in a timely manner. The 2000 census, for example, was published in 49 languages. Some of the languages, including German and French, were predictable. Some of them, such as Chamarro, Dinka, and Ilocano, are less well known. The volume of data has required the bureau to automate the process as much as possible. During the decennial census the bureau makes an electronic image of every questionnaire that is returned. All returned envelopes are automatically sorted, and households that did not comply are automatically identified so that a census worker can interview the residents. The information collected is fed through data-analysis software that provides a statistical snapshot of the United States that is as complete as possible.

In addition to the decennial census, the Census Bureau manages a number of surveys and performs numerous statistical analyses. The bureau publishes a Census of Manufacturers, an American Housing Survey, a Consumer Expenditure Survey, a Survey of Income and Program Participation, and numerous other surveys of interest to economic planners. It is important to realize that each such activity is an application of statistics. The Census Bureau collects massive amounts of data and provides statistical analyses of the data it obtains. The bureau does not use this information; it is not a policy arm of the government. Instead, it provides information to the many interested governmental and private institutions that need it as an aid in their decision-making processes. This is one of the main reasons statistics are valuable.

Political Polling

The polling industry is an important application of the ideas described in this book. Opinion polls are a part of daily life. They often form the central topic of news shows. The question, Are you better off since the current president took office? has been asked of likely voters for many years by people running for office. This question is frequently used in polls commissioned by incumbents as well. The commissioning of polls is a very large business, and the results of the polls affect everything from commercial advertising to the political discourse of nations. It has not always been so. In the 19th century in the United States there were only the most informal methods of sampling opinion. During election years some hotels would provide a space on the hotel registry for guests to express their presidential preferences. Newspapers occasionally sampled public opinion by sending out reporters to ask questions of the "general public." There was nothing scientific about the sampling methods employed by these news organizations—and there was nothing very impressive about the accuracy of the surveys, either.

The first scientific use of a political poll occurred in Iowa in 1932. Viola "Ola" Babcock Miller, a leader in the suffrage movement, was running for secretary of state of Iowa. (She was successful in her bid to become Iowa's first female secretary of state.) While she was running, her son-in-law approached her about the possibility of conducting a political poll for her campaign. As a Ph.D. student he had developed a method of surveying readership of newspaper stories and advertisements, and he was eager to test his ideas by forecasting a political contest. Miller agreed to allow the test, and the poll correctly indicated that she would win. The name of Miller's son-in-law was George H. Gallup (1901–84), founder of the Gallup poll. The era of public opinion research had begun.

Not everyone was quick to notice the importance of Gallup's ideas, but Gallup himself was confident of his ability to predict the behavior of large groups of people on the basis of the analysis of small samples. In the 1936 presidential election he sought to obtain further exposure for his ideas on scientific polling.

President Franklin D. Roosevelt was running for reelection against Alfred M. Landon, the Republican challenger. The magazine *Literary Digest* had gained some fame for accurate straw polls in elections preceding the 1936 contest. (A straw poll acquires its data by using nonscientific methods.) The *Digest* worked under the belief that the bigger the sample, the more accurate the results, and to that end it sent out 10 million questionnaires. It received 2 million responses. Of course, the respondents were not randomly selected. No matter what procedure the *Digest* used to obtain addresses, the respondents were *self-selected* in the sense that only those who cared enough about the poll to register their opinions had their opinions registered. The *Digest's* straw poll indicated a convincing win for Landon.

Young Gallup, fresh from his success in his mother-in-law's campaign, launched his own survey of attitudes about the 1936 presidential election. His results, based on a much smaller but scientifically selected sample, indicated that Roosevelt would win the election with 55.7 percent of the vote. Despite the prestige of the *Digest*, Gallup made his prediction with as much fanfare as he could generate. The result of the election was a landslide win for Roosevelt, who garnered 62.5 percent of the vote. By contemporary standards Gallup's approximately 7 percent error left a lot to be desired, but by the standards of the time it was a remarkable achievement. Within four years President Roosevelt was commissioning his own polls to track American attitudes about the progress of the Second World War.

In the years since Roosevelt's reelection over Landon, polling has become increasingly important. Initially, polling was used to reveal public attitudes and behavior: How would a particular segment of the population vote? What marketing strategy would sell the most cigarettes? What do people look for when buying a car? If polling can predict the outcome of elections, then it can be used to devise commercial marketing strategies as well. From a mathematical point of view there is no difference, and this has been a source of concern for some. The former presidential candidate Adlai Stevenson remarked, "The idea that you can merchandise candidates for high office like breakfast cereal is the ultimate indignity to the democratic process," and, of course, this is precisely what polling enables one to do. Merchandising political candidates with the same techniques used to merchandise breakfast cereal not only can happen, it has become a standard part of the process in both state and federal elections.

The accuracy, and hence the value, of carefully conducted polls has on occasion been called into question. One of the most spectacular failures of political polling techniques involved the 1948 presidential election in which President Harry Truman was opposed by Governor Thomas Dewey. Many people considered Truman the underdog in the election. In fact, polls taken seven

In what is probably the most famous public opinion polling error in the history of U.S. presidential elections, Dewey was widely forecast to win over Truman in the 1948 election. (©1948 The Washington Post. Reprinted with permission. U.S. Senate Collection, Center for Legislative Archives)

months before the election indicated that he had only a 34 percent approval rating. During the election campaign, polling organizations used something called quota sampling to obtain what they hoped would be a representative sample of the electorate. Quota sampling generally involves census data. A population of interest—in this case, prospective voters—is broken down into various groups, for example, women, men, and people older than age 65, and each group is sampled separately. The difficult part involves identifying the population of likely voters. After the population is segmented, interviewers are allowed to choose individuals from each subgroup. Leaving the choice of individuals to the interviewer means the poll designer loses control over the process. Individual choices can lead to individual biases, and such was the case during the Truman–Dewey campaign. On the eve of the election, Dewey was widely predicted to win by a landslide, but it was Truman who won the election.

Nor did all the failures of political polling occur 50 years ago. The 2000 presidential elections and the 2002 midterm elections are both examples of U.S. elections that had outcomes that were not clearly foreseen by the political polls. Accounting for an inaccurate forecast is an inexact science. Sometimes, bad forecasting is explained away by blaming the electorate. A *volatile electorate*—that is, a significant percentage of people who decide how to cast their votes at the last minute—is often identified as the culprit. A good poll, however, should be able to identify "volatility" *before* the election instead of afterward. Alternatively, some believe that the poll itself can influence the outcome of an election. This is the so-called bandwagon effect. The theory is that once people believe they know the winner, they are more likely to vote for the winner. George Gallup searched for this effect in past elections, but he dismissed the idea that the bandwagon effect played a significant role.

Mathematically, there are several difficulties that must be overcome in order to obtain reliable forecasts. The first difficulty involves obtaining a clear definition of the set of interest. This is not always easy: The characteristics of the parent set, or what Deming called the universe, are not known. Who, for example, belongs to the set of "likely" voters? Is it the set of all people who voted in two of the last three presidential elections? This criterion cannot be applied even to registered voters younger than age 25 because, depending on when the survey is taken, they could not have been registered long enough to vote in the previous two presidential elections. Furthermore, some elections simply attract more interest than others. The set of likely voters can vary significantly from one election to the next. Identifying likely voters is not easy.

Another problem arises when one tries to obtain a representative sample of the set of all likely voters. In theory, the best sample is a randomly drawn sample, but achieving a randomly drawn sample turns out to be quite difficult. Many people, for example, simply will not answer polling questions. One reason is fatigue. During an election cycle many people are contacted by phone at home multiple times—and often during supper—by interviewers who are attempting to collect data for a variety of surveys. Eventually, many people stop cooperating. Noncompliance is a serious problem in attempting to obtain a random sample because noncompliance may be correlated with voting preferences. When this is true, certain voting preferences are underrepresented in the sample. Some analysts believe that during the 2002 midterm elections Republican voters were more likely to use call screening technology to weed out phone surveys, and that this was one reason that Republican turnout was higher than predicted. Can we be sure that this explanation is correct? No. Unless they pick up their phones and answer, or they respond to a mail survey, or researchers find some other economical method of surveying their preferences, it is very difficult to prove or disprove the truth of the statement. The insights about the accuracy and economics of sampling contained in Deming's book *Some Theory of Sampling* are as valid today as they were a half-century ago.

The "luck of the draw" is a common theme in literature. We are subject to the vagaries of "fortune," whether that is expressed in good weather or bad, sickness or health. Even the genes that, in part, make us who we are were inherited at random from our parents: Given a set of genes, the effect of the set on the phenotype of the organism is often predictable. However, given the parents of the organism, predicting which genes will be inherited requires probability (as a general rule). Although random phenomena are important to all of us—they are some of the principal factors that determine how we live our lives—for the better part of humanity's 5,000-year recorded history people simply endured and wondered about what "fate" had in store for them.

Less than 400 years ago something remarkable happened. People began to develop the conceptual tools necessary to understand randomness and uncertainty. The value of the work of Pascal and Fermat in probability and of Graunt and Halley in statistics was quickly recognized by their peers, and the race to understand the nature of chance and unpredictable variation began.

During the intervening years, the ideas and language of probability and statistics have become a part of our everyday experience. News reports, weather reports, and sports facts—everything that we are accustomed to classifying as "news"—are now regularly expressed in the language of statistics and probability. No other branch of mathematics has had such success in capturing the popular imagination. No other branch of mathematics has proved so useful for expressing the ideas that are important to so many people. Probability is used to estimate the safety of everything from vaccines to bridges; statistics is used to help formulate public policy and even to tell us what we as a people are thinking and doing. Probability and statistics now lie at the heart of the way we understand the world.

Despite the utility of this type of mathematics no mathematician would assert that we have done more than scratch the surface of these remarkable disciplines. As our appreciation for the interconnectivity of complex systems increases, the need for increasingly sophisticated statistical techniques to analyze the data sets that purport to describe these systems is keenly felt by all researchers. New uses for probability continue to be discovered even as mathematicians continue to debate the connections between the mathematical discipline of probability and its real-world applications.

The history of probability and statistics has just begun.

CHRONOLOGY

ca. 3000 B.C.E.
Hieroglyphic numerals are in use in Egypt.

ca. 2500 B.C.E.
Construction of the Great Pyramid of Khufu takes place.

ca. 2400 B.C.E.
An almost complete system of positional notation is in use in Mesopotamia.

ca. 1800 B.C.E.
The Code of Hammurabi is promulgated.

ca. 1650 B.C.E.
The Egyptian scribe Ahmes copies what is now known as the Ahmes (or Rhind) papyrus from an earlier version of the same document.

ca. 1200 B.C.E.
The Trojan War is fought.

ca. 740 B.C.E.
Homer composes the *Odyssey* and the *Iliad*, his epic poems about the Trojan War.

ca. 585 B.C.E.
Thales of Miletus carries out his research into geometry, marking the beginning of mathematics as a deductive science.

ca. 540 B.C.E.
Pythagoras of Samos establishes the Pythagorean school of philosophy.

ca. 500 B.C.E.
Rod numerals are in use in China.

ca. 420 B.C.E.
Zeno of Elea proposes his philosophical paradoxes.

ca. 399 B.C.E.
Socrates dies.

ca. 360 B.C.E.
Eudoxus, author of the method of exhaustion, carries out his research into mathematics.

ca. 350 B.C.E.
The Greek mathematician Menaechmus writes an important work on conic sections.

ca. 347 B.C.E.
Plato dies.

332 B.C.E.
Alexandria, Egypt, center of Greek mathematics, is established.

ca. 300 B.C.E.
Euclid of Alexandria writes *Elements*, one of the most influential mathematics books of all time.

ca. 260 B.C.E.
Aristarchus of Samos discovers a method for computing the ratio of the Earth–Moon distance to the Earth–Sun distance.

ca. 230 B.C.E.
Eratosthenes of Cyrene computes the circumference of Earth.

Apollonius of Perga writes *Conics*.

Archimedes of Syracuse writes *The Method, Equilibrium of Planes*, and other works.

206 B.C.E.
The Han dynasty is established; Chinese mathematics flourishes.

ca. A.D. 150
Ptolemy of Alexandria writes *Almagest*, the most influential astronomy text of antiquity.

ca. A.D. 250
Diophantus of Alexandria writes *Arithmetica*, an important step forward for algebra.

ca. 320
Pappus of Alexandria writes his *Collection*, one of the last influential Greek mathematical treatises.

415
The death of the Alexandrian philosopher and mathematician Hypatia marks the end of the Greek mathematical tradition.

ca. 476
The astronomer and mathematician Aryabhata is born; Indian mathematics flourishes.

ca. 630
The Hindu mathematician and astronomer Brahmagupta writes *Brahma-sphuta-siddhānta*, which contains a description of place-value notation.

641
The Library of Alexandria is burned.

ca. 775
Scholars in Baghdad begin to translate Hindu and Greek works into Arabic.

ca. 830
Mohammed ibn-Mūsā al-Khwārizmī writes *Hisāb al-jabr wa'l muqābala*, a new approach to algebra.

833
Al-Ma'mūn, founder of the House of Wisdom in Baghdad, Iraq, dies.

ca. 840
The Jainist mathematician Mahavira writes *Ganita Sara Samgraha*, an important mathematical textbook.

1123
Omar Khayyam, author of *Al-jabr w'al muqābala* and the *Rubáiyát*, the last great classical Islamic mathematician, dies.

ca. 1144
Bhaskara II writes the *Lilavati* and the *Vija-Ganita*, two of the last great works in the classical Indian mathematical tradition.

1071
William the Conqueror quells the last of the English rebellions.

1086
An intensive survey of the wealth of England is carried out and summarized in the tables and lists of the *Domesday Book*.

ca. 1202
Leonardo of Pisa (Fibonacci), author of *Liber abaci*, arrives in Europe.

1360
Nicholas Oresme, a French mathematician and Roman Catholic bishop, represents distance as the area beneath a velocity line.

1471
The German artist Albrecht Dürer is born.

1482
Leonardo da Vinci begins his diaries.

ca. 1541
Niccolò Fontana, an Italian mathematician also known as Tartaglia, discovers a general method for factoring third-degree algebraic equations.

1543
Copernicus publishes *De Revolutionibus*, marking the start of the Copernican Revolution.

1545
Girolamo Cardano, an Italian mathematician and physician, publishes *Ars Magna*, marking the beginning of modern algebra. Later he publishes *Liber de Ludo Aleae*, the first book on probability.

ca. 1554
Sir Walter Raleigh, the explorer, adventurer, amateur mathematician, and patron of the mathematician Thomas Harriot, is born.

1579

François Viète, a French mathematician, publishes *Canon Mathematicus*, marking the beginning of modern algebraic notation.

1585

The Dutch mathematician and engineer Simon Stevin publishes "La disme."

1609

Johannes Kepler, the proponent of Kepler's laws of planetary motion, publishes *Astronomia Nova*.

Galileo Galilei begins his astronomical observations.

1621

The English mathematician and astronomer Thomas Harriot dies. His only work, *Artis Analyticae Praxis*, is published in 1631.

ca. 1630

The French lawyer and mathematician Pierre de Fermat begins a lifetime of mathematical research. He is the first person to claim to have proved "Fermat's last theorem."

1636

Gérard (Girard) Desargues, a French mathematician and engineer, publishes *Traité de la section perspective*, which marks the beginning of projective geometry.

1637

René Descartes, a French philosopher and mathematician, publishes *Discours de la méthode*, permanently changing both algebra and geometry.

1638

Galileo Galilei publishes *Dialogues Concerning Two New Sciences* while under arrest.

1640

Blaise Pascal, a French philosopher, scientist, and mathematician, publishes *Essaie sur les coniques*, an extension of the work of Desargues.

1642

Blaise Pascal manufactures an early mechanical calculator, the Pascaline.

1648

The Thirty Years' War, a series of conflicts that involved much of Europe, ends.

1649

Oliver Cromwell takes control of the English government after a civil war.

1654

Pierre de Fermat and Blaise Pascal exchange a series of letters about probability, thereby inspiring many mathematicians to study the subject.

1655

John Wallis, an English mathematician and clergyman, publishes *Arithmetica Infinitorum*, an important work that presages calculus.

1657

Christian Huygens, a Dutch mathematician, astronomer, and physicist, publishes *Ratiociniis in Ludo Aleae*, a highly influential text in probability theory.

1662

John Graunt, an English businessman and a pioneer in statistics, publishes his research on the London Bills of Mortality.

1673

Gottfried Leibniz, a German philosopher and mathematician, constructs a mechanical calculator that can perform addition, subtraction, multiplication, division, and extraction of roots.

1683

Seki Kowa, a Japanese mathematician, discovers the theory of determinants.

1684

Gottfried Leibniz publishes the first paper on calculus, *Nova Methodus pro Maximis et Minimis*.

1687

Isaac Newton, a British mathematician and physicist, publishes *Philosophiae Naturalis Principia Mathematica*, beginning a new era in science.

1693

Edmund Halley, a British mathematician and astronomer, undertakes a statistical study of the mortality rate in Breslau, Germany.

1698

Thomas Savery, an English engineer and inventor, patents the first steam engine.

1705

Jacob Bernoulli, a Swiss mathematician, dies. His major work on probability, *Ars Conjectandi*, is published in 1713.

1712

The first Newcomen steam engine is installed.

1718

Abraham de Moivre, a French mathematician, publishes *The Doctrine of Chances*, the most advanced text of the time on the theory of probability.

1743

The Anglo-Irish Anglican bishop and philosopher George Berkeley publishes *The Analyst*, an attack on the new mathematics pioneered by Isaac Newton and Gottfried Leibniz.

The French mathematician and philosopher Jean le Rond d'Alembert begins work on the *Encyclopédie*, one of the great works of the Enlightenment.

1748

Leonhard Euler, a Swiss mathematician, publishes his *Introductio*.

1749

The French mathematician and scientist George-Louis Leclerc Buffon publishes the first volume of *Histoire naturelle*.

1750

Gabriel Cramer, a Swiss mathematician, publishes "Cramer's rule," a procedure for solving systems of linear equations.

1760

Daniel Bernoulli, a Swiss mathematician and scientist, publishes his probabilistic analysis of the risks and benefits of variolation against smallpox.

1761

Thomas Bayes, an English theologian and mathematician, dies. His "Essay Towards Solving a Problem in the Doctrine of Chances" is published two years later.

The English scientist Joseph Black proposes the idea of latent heat.

1762

Catherine II (Catherine the Great) is proclaimed empress of Russia.

1769

James Watt obtains his first steam engine patent.

1775

American colonists and British troops fight battles at Lexington and Concord, Massachusetts.

1778

Voltaire (François-Marie Arouet), a French writer and philosopher, dies.

1781

William Herschel, a German-born British musician and astronomer, discovers Uranus.

1789

Unrest in France culminates in the French Revolution.

1793

The Reign of Terror, a period of brutal, state-sanctioned repression, begins in France.

1794

The French mathematician Adrien-Marie Legendre (or Le Gendre) publishes his *Éléments de géométrie*, a text that will influence mathematics education for decades.

Antoine-Laurent Lavoisier, a French scientist and discoverer of the law of conservation of matter, is executed by the French government.

1798

Benjamin Thompson (Count Rumford), a British physicist, proposes the equivalence of heat and work.

1799

Napoléon seizes control of the French government.

Caspar Wessel, a Norwegian mathematician and surveyor, publishes the first geometric representation of the complex numbers.

1801

Carl Friedrich Gauss, a German mathematician, publishes *Disquisitiones Arithmeticae*.

1805

Adrien-Marie Legendre, a French mathematician, publishes "Nouvelles methodes pour la determination des orbietes des comets," which contains the first description of the method of least squares.

1806

Jean-Robert Argand, a French bookkeeper, accountant, and mathematician, develops the Argand diagram to represent complex numbers.

1812

Pierre-Simon Laplace, a French mathematician, publishes *Theorie analytique des probabilities*, the most influential 19th-century work on the theory of probability.

1815

Napoléon suffers final defeat at the battle of Waterloo.

Jean-Victor Poncelet, a French mathematician and the "father of projective geometry," publishes *Traité des propriétés projectives des figures.*

1824

The French engineer Sadi Carnot publishes *Réflections*, wherein he describes the Carnot engine.

Niels Henrik Abel, a Norwegian mathematician, publishes his proof of the impossibility of algebraically solving a general fifth-degree equation.

1826

Nikolay Ivanovich Lobachevsky, a Russian mathematician and "the Copernicus of geometry," announces his theory of non-Euclidean geometry.

1828

Robert Brown, a Scottish botanist, publishes the first description of Brownian motion in "A Brief Account of Microscopical Observations."

1830

Charles Babbage, a British mathematician and inventor, begins work on his analytical engine, the first attempt at a modern computer.

1832

Janos Bolyai, a Hungarian mathematician, publishes *Absolute Science of Space.*

The French mathematician Evariste Galois is killed in a duel.

1843

James Prescott Joule publishes his measurement of the mechanical equivalent of heat.

1846

The planet Neptune is discovered by the French mathematician Urbain-Jean-Joseph Le Verrier through a mathematical analysis of the orbit of Uranus.

1847

Georg Christian von Staudt publishes *Geometrie der Lage*, which shows that projective geometry can be expressed without any concept of length.

1848

Bernhard Bolzano, a Czech mathematician and theologian, dies. His study of infinite sets, *Paradoxien des Unendlichen*, is published for the first time in 1851.

1850

Rudolph Clausius, a German mathematician and physicist, publishes his first paper on the theory of heat.

1851

William Thomson (Lord Kelvin), a British scientist, publishes "On the Dynamical Theory of Heat."

1854

George Boole, a British mathematician, publishes *Laws of Thought*. The mathematics contained therein will make possible the design of computer logic circuits.

The German mathematician Bernhard Riemann gives the historic lecture "On the Hypotheses That Form the Foundations of Geometry." The ideas therein will play an integral part in the theory of relativity.

1855

John Snow, a British physician, publishes "On the Mode of Communication of Cholera," the first successful epidemiological study of a disease.

1859

James Clerk Maxwell, a British physicist, proposes a probabilistic model for the distribution of molecular velocities in a gas.

Charles Darwin, a British biologist, publishes *On the Origin of Species by Means of Natural Selection*.

1861

The American Civil War begins.

1866

The Austrian biologist and monk Gregor Mendel publishes his ideas on the theory of heredity in "Versuche über Pflanzen-hybriden."

1867

The Canadian Articles of Confederation unify the British colonies of North America.

1871

Otto von Bismarck is appointed first chancellor of the German Empire.

1872

The German mathematician Felix Klein announces his Erlanger Programm, an attempt to categorize all geometries with the use of group theory.

Lord Kelvin (William Thomson) develops an early analog computer to predict tides.

Richard Dedekind, a German mathematician, rigorously establishes the connection between real numbers and the real number line.

1874

Georg Cantor, a German mathematician, publishes "Über eine Eigenschaft des Inbegriffes aller reelen algebraischen Zahlen," a pioneering paper that shows that not all infinite sets are the same size.

1890

The Hollerith tabulator, an important innovation in calculating machines, is installed by the United States Census for use in the 1890 census.

1899

The German mathematician David Hilbert publishes the definitive axiomatic treatment of Euclidean geometry.

1900

David Hilbert announces his list of mathematics problems for the 20th century.

The Russian mathematician Andrey Andreyevich Markov begins his research into the theory of probability.

1901
Henri-Léon Lebesgue, a French mathematician, develops his theory of integration.

1905
Ernst Zermelo, a German mathematician, undertakes the task of axiomatizing set theory.

Albert Einstein, a German-born American physicist, begins to publish his discoveries in physics.

1906
Marian Smoluchowski, a Polish scientist, publishes his insights into Brownian motion.

1908
The Hardy-Weinberg law, containing ideas fundamental to population genetics, is published.

1910
Bertrand Russell, a British logician and philosopher, and Alfred North Whitehead, a British mathematician and philosopher, publish *Principia Mathematica*, an important work on the foundations of mathematics.

1914
World War I begins.

1917
Vladimir Ilyich Lenin leads a revolution that results in the founding of the Union of Soviet Socialist Republics.

1918
World War I ends.

The German mathematician Emmy Noether presents her ideas on the role of symmetries in physics.

1929

Andrey Nikolayevich Kolmogorov, a Russian mathematician, publishes *General Theory of Measure and Probability Theory*, putting the theory of probability on a firm axiomatic basis for the first time.

1930

Ronald Aylmer Fisher, a British geneticist and statistician, publishes *Genetical Theory of Natural Selection*, an important early attempt to express the theory of natural selection in mathematics.

1931

Kurt Gödel, an Austrian-born American mathematician, publishes his incompleteness proof.

The Differential Analyzer, an important development in analog computers, is developed at Massachusetts Institute of Technology.

1933

Karl Pearson, a British innovator in statistics, retires from University College, London.

1935

George Horace Gallup, a U.S. statistician, founds the American Institute of Public Opinion.

1937

The British mathematician Alan Turing publishes his insights on the limits of computability.

1939

World War II begins.

William Edwards Deming joins the United States Census Bureau.

1945

World War II ends.

1946

The Electronic Numerical Integrator and Calculator (ENIAC) computer begins operation at the University of Pennsylvania.

1948
While working at Bell Telephone Labs in the United States, Claude Shannon publishes "A Mathematical Theory of Communication," marking the beginning of the Information Age.

1951
The Universal Automatic Computer (UNIVAC I) is installed at the U.S. Bureau of the Census.

1954
FORTRAN (FORmula TRANslator), one of the first high-level computer languages, is introduced.

1956
The American Walter Shewhart, innovator in the field of quality control, retires from Bell Telephone Laboratories.

1957
Olga Oleinik publishes "Discontinuous solutions to nonlinear differential equations," a milestone in mathematical physics.

1964
IBM Corporation introduces the IBM System/360 computer for government agencies and large businesses.

1965
Andrey Nikolayevich Kolmogorov establishes the branch of mathematics now known as Kolmogorov complexity.

1966
APL (A Programming Language) is implemented on the IBM System/360 computer.

1972
Amid much fanfare, the French mathematician and philosopher René Thom establishes a new field of mathematics called catastrophe theory.

1973
The C computer language, developed at Bell Laboratories, is essentially completed.

1975

The French geophysicist Jean Morlet helps develop a new kind of analysis based on what he calls "wavelets."

1977

Digital Equipment Corporation introduces the VAX computer.

1981

IBM Corporation introduces the IBM personal computer (PC).

1989

The Belgian mathematician Ingrid Daubechies develops what has become the mathematical foundation for today's wavelet research.

1991

The Union of Soviet Socialist Republics dissolves into 15 separate nations.

1995

The British mathematician Andrew Wiles publishes the first proof of Fermat's last theorem.

Cray Research introduces the CRAY E-1200, a machine that sustains a rate of one terraflop (1 trillion calculations per second) on real-world applications.

The JAVA computer language is introduced commercially by Sun Microsystems.

1997

René Thom declares the mathematical field of catastrophe theory "dead."

2002

Experimental Mathematics celebrates its 10th anniversary. It is a refereed journal dedicated to the experimental aspects of mathematical research.

Manindra Agrawal, Neeraj Kayal, and Nitin Saxena create a brief, elegant algorithm to test whether a number is prime, thereby solving an important centuries-old problem.

2003
Grigory Perelman produces what may be the first complete proof of the Poincaré conjecture, a statement on the most fundamental properties of three-dimensional shapes.

GLOSSARY

assignable cause variation   variation in the quality of a product or process that is due to nonrandom factors

axiom   a statement accepted as true to serve as a basis for deductive reasoning. Today the words *axiom* and *postulate* are synonyms

Bayesian   of or relating to that part of the theory of probability concerned with estimating probabilities through the use of prior knowledge

Bayes's theorem   the first theorem on conditional probabilities. If one knows the probability of event *A* given that event *B* has already occurred and certain auxiliary information, Bayes's theorem allows one to compute the probability of event *B* given that event *A* is known to have occurred. Bayes's theorem marks the beginning of the study of inverse probability

Brownian motion   random motion of microscopic particles immersed in a liquid or gas that is due to impacts of the surrounding molecules

chance cause variation   variation in quality of a product or process due to random factors

control chart   a statistical tool designed to help measure the degree of economic control exerted over an industrial process; also known as a Shewhart control chart

curve fitting   any of several mathematical methods for determining which curve—chosen from a well-defined set of curves—best represents a data set

determinism   in science, the philosophical principle that future and past states of a system can be predicted from certain equations and knowledge of the present state of the system

economic control cost-efficient management of a process

epidemiology the branch of medicine that studies the distribution of disease in human populations. Statistics is one of the principal investigative tools of researchers in this field of science

frequentist of or relating to that part of the theory of probability concerned with estimating probabilities by using the measured frequencies of previous outcomes

independent not influenced by another event. In probability two events are independent of one another if the occurrence or nonoccurrence of one event has no effect on the probability of occurrence or nonoccurrence of the other event

information theory a statistical theory of information that provides a set of methods for measuring the amount of information present in a message and the efficiency with which the message is encoded

inverse probability the concept of probability that arose out of Bayes's theorem

law of large numbers a theorem that asserts that over the course of many trials the frequency with which any particular event occurs approaches the probability of the event

Markov chain a random process consisting of a set of discrete states or a chain of events in which the probability of future states does not depend on the occurrence or nonoccurrence of past states

mean the center or average of a set of measurements

measure theory a branch of mathematics that generalizes the problems of measuring length, area, and volume to the more general problem of measuring the space occupied by arbitrary sets of points

normal curve a curve used to estimate the probability of occurrence of events for many common random processes; also known as the bell curve or normal distribution

normal distribution see NORMAL CURVE

Poisson distribution   a curve used to estimate the probability of events for certain types of random processes

probability   the branch of mathematics concerned with the study of chance

random pattern   a collection of events whose outcomes could not have been known before their occurrence

representative sample   a segment of a larger population whose properties reflect the statistical structure of the larger, parent set from which it was drawn

statistics   the branch of mathematics dedicated to the collection and analysis of data

tests of significance   a collection of statistical methods for determining whether observed variation in a sample represents chance variation that always occurs when random samples are drawn from a parent population or whether the observed variation is due to nonrandom causes

universe   in sampling theory, the set that contains all elements relevant to a particular statistical analysis

variance   a measure of the variation about the mean in a set of measurements

FURTHER READING

MODERN WORKS

Best, Joel. *Damned Lies and Statistics: Untangling Numbers from the Media, Politicians, and Activists*. Berkeley: University of California Press, 2001. A critical and creative examination of the uses and abuses of statistics and statistical reasoning.

Borel, Émile. *Probabilities and Life*. New York: Dover Publications, 1962. A short, carefully written introduction to probability and its applications.

Boyer, Carl B., and Uta C. Merzbach. *A History of Mathematics*. New York: John Wiley & Sons, 1991. Boyer was one of the preeminent mathematics historians of the 20th century. This work contains much interesting biographical information. The mathematical information assumes a fairly strong mathematical background.

Bruno, Leonard C. *Math and Mathematicians: The History of Mathematics Discoveries around the World*, 2 vols. Detroit, Mich.: U.X.L, 1999. Despite its name there is little mathematics in this two-volume set. What you will find is a very large number of brief biographies of many individuals who are important in the history of mathematics.

Courant, Richard, and Herbert Robbins. *What Is Mathematics? An Elementary Approach to Ideas and Mathematics*. New York: Oxford University Press, 1941. A classic and exhaustive answer to the question posed in the title. Courant was an important and influential 20th-century mathematician.

Cushman, Jean. *Do You Want to Bet? Your Chance to Find Out about Probability*. New York: Clarion Books, 1991. A simple and sometimes silly introduction to some of the basic concepts of probability—this is still a good place to begin.

David, Florence N. *Games, Gods and Gambling: A History of Probability and Statistical Ideas.* New York: Dover Publications, 1998. This is an excellent account of the early history of probability and statistics. In addition to an analysis of the early history of the subject, this book also contains Galileo's writings on probability, Fermat's and Pascal's correspondence on probability, and a brief excerpt from Abraham de Moivre's *Doctrine of Chances.* Highly recommended.

Dewdney, Alexander K. *200% of Nothing: An Eye-Opening Tour through the Twists and Turns of Math Abuse and Innumeracy.* New York: John Wiley & Sons, 1993. A critical look at how mathematical reasoning has been abused to distort truth.

Eastaway, Robert, and Jeremy Wyndham. *Why Do Buses Come in Threes? The Hidden Mathematics of Everyday Life.* New York: John Wiley & Sons, 1998. Nineteen lighthearted essays on the mathematics underlying everything from luck to scheduling problems.

Eves, Howard. *An Introduction to the History of Mathematics.* New York: Holt, Rinehart & Winston, 1953. This well-written history of mathematics places special emphasis on early mathematics. It is unusual because the history is accompanied by numerous mathematical problems. (The solutions are in the back of the book.)

Freudenthal, Hans. *Mathematics Observed.* New York: McGraw-Hill, 1967. A collection of seven survey articles about math topics from computability to geometry to physics (some more technical than others).

Gardner, Martin. *The Colossal Book of Mathematics.* New York: Norton, 2001. Martin Gardner had a gift for seeing things mathematically. This "colossal" book contains sections on geometry, algebra, probability, logic, and more.

Gardner, Martin. *Order and Surprise.* Buffalo, N.Y.: Prometheus Books, 1983. A worthwhile contribution to the subject of probability in a highly readable form.

Gigerenzer, Gerd. *Calculated Risks: How to Know When Numbers Deceive You.* New York: Simon & Schuster, 2002. A fascinating look at how mathematics is used to gain insight into everything from breast cancer screening to acquired immunodeficiency syndrome (AIDS) counseling to deoxyribonucleic acid (DNA) fingerprinting.

Guillen, Michael. *Bridges to Infinity: The Human Side of Mathematics.* Los Angeles: Jeremy P. Tarcher, 1983. This book consists of an engaging nontechnical set of essays on mathematical topics, including non-Euclidean geometry, transfinite numbers, and catastrophe theory.

Hoffman, Paul. *Archimedes' Revenge: The Joys and Perils of Mathematics.* New York: Ballantine, 1989. A relaxed, sometimes silly look at an interesting and diverse set of math problems ranging from prime numbers and cryptography to Turing machines and the mathematics of democratic processes.

Kline, Morris. *Mathematics for the Nonmathematician.* New York: Dover Publications, 1985. An articulate, not very technical overview of many important mathematical ideas.

Kline, Morris. *Mathematics in Western Culture.* New York: Oxford University Press, 1953. An excellent overview of the development of Western mathematics in its cultural context, this book is aimed at an audience with a firm grasp of high-school-level mathematics.

Nahin, Paul J. *Dueling Idiots and Other Probability Puzzlers.* Princeton, N.J.: Princeton University Press, 2000. This is a collection of entertaining "puzzlers" analyzed from a mathematical perspective.

Orkin, Michael. *Can You Win? The Real Odds for Casino Gambling, Sports Betting, and Lotteries.* New York: W. H. Freeman, 1991. An enlightening, updated look at the first of all applications of the theory of probability.

Packel, Edward W. *The Mathematics of Games and Gambling.* Washington, D.C.: Mathematical Association of America, 1981. A good introduction to the mathematics underlying two of humankind's oldest forms of recreation and disappointment.

Pappas, Theoni. *The Joy of Mathematics.* San Carlos, Calif.: World Wide/Tetra, 1986. Aimed at a younger audience, this work searches for interesting applications of mathematics in the world around us.

Pierce, John R. *An Introduction to Information Theory: Symbols, Signals and Noise.* New York: Dover Publications, 1961. Despite the sound of the title, this is not a textbook. Among other topics, Pierce, formerly of Bell Laboratories, describes some of the mathematics involved in measuring the amount of information present in a text—an important application of probability theory.

Salsburg, David. *The Lady Tasting Tea: How Statistics Revolutionized Science in the Twentieth Century.* New York: W. H. Freeman, 2001. A very detailed look at the history of statistics and statistical thinking in the 20th century.

Sawyer, Walter W. *What Is Calculus About?* New York: Random House, 1961. A highly readable description of a sometimes-intimidating, historically important subject. Absolutely no calculus background is required.

Schiffer, M., and Leon Bowden. *The Role of Mathematics in Science.* Washington, D.C.: Mathematical Association of America, 1984. The first few chapters of this book, ostensibly written for high school students, will be accessible to many students; the last few chapters will find a much narrower audience.

Stewart, Ian. *From Here to Infinity.* New York: Oxford University Press, 1996. A well-written, very readable overview of several important contemporary ideas in geometry, algebra, computability, chaos, and mathematics in nature.

Swetz, Frank J., editor. *From Five Fingers to Infinity: A Journey through the History of Mathematics.* Chicago: Open Court, 1994. This is a fascinating, though not especially focused, look at the history of mathematics. Highly recommended.

Tabak, John. *Math and the Laws of Nature. History of Mathematics.* New York: Facts On File, 2004. More information about the relationships that exist between random processes and the laws of nature.

Thomas, David A. *Math Projects for Young Scientists.* New York: Franklin Watts, 1988. This project-oriented text gives an introduction to several historically important geometry problems.

Weaver, Jefferson H. *What Are the Odds? The Chances of Extraordinary Events in Everyday Life.* Amherst, N.Y.: Prometheus, 2001. A lighthearted treatment of probability and statistics and their applications to romance, death, war, and the chance of becoming a rock star.

ORIGINAL SOURCES

It can sometimes deepen our appreciation of an important mathematical discovery to read the discoverer's own description. Often this is not possible, because the description is too technical. Fortunately, there are exceptions. Sometimes the discovery is accessible because the idea does not require a lot of technical background to appreciate it. Sometimes, the discoverer writes a nontechnical account of the technical idea that she or he has discovered. Here are some classic papers:

Bernoulli, J. The Law of Large Numbers. In *The World of Mathematics.* Vol. 3, edited by James Newman. New York: Dover Publications, 1956. This excerpt contains Jacob Bernoulli's own description of one of the great discoveries in the history of probability and statistics.

Fermat, Pierre de, and Pascal, Blaise. The exchange of letters between Pierre de Fermat and Blaise Pascal marks the beginning of the modern theory of probability. These letters have been translated and appear as an appendix in the book *Games, Gods and Gambling: A History of Probability and Statistical Ideas* by Florence N. David (New York: Dover Publications, 1998).

Fisher, R. A. Mathematics of a Lady Tasting Tea. In *The World of Mathematics.* Vol. 3, edited by James R. Newman. New York: Dover Publications, 1956. A wonderful, largely nontechnical account of the challenges involved in designing an experiment to test a hypothesis.

Graunt, J. Foundations of Vital Statistics. In *The World of Mathematics*. Vol. 3, edited by James R. Newman. New York: Dover Publications, 1956. This excerpt from a 1662 paper marks the beginning of the modern theory of statistics. It is still remarkable for its clarity of thought and careful analysis.

Halley, Edmund. First Life Insurance Tables. In *The World of Mathematics*. Vol. 3, edited by James R. Newman. New York: Dover Publications, 1956. Part of Edmund Halley's groundbreaking statistical survey of the bills of mortality of Breslau.

Hardy, Godfrey H. *A Mathematician's Apology*. Cambridge, U.K.: Cambridge University Press, 1940. Hardy was an excellent mathematician and a good writer. In this oft-quoted and very brief book Hardy seeks to explain and sometimes justify his life as a mathematician.

Laplace, P. Concerning Probability. In *The World of Mathematics*. Vol. 2, edited by James R. Newman. New York: Dover Publications, 1956. A nontechnical introduction to some fundamental ideas in the field of probability by one of the great innovators in the field.

INTERNET RESOURCES

Athena Earth and Space Science for K–12. Available on-line. URL: http://inspire.ospi.wednet.edu:8001/. Updated May 13, 1999. Funded by the National Aeronautics and Space Administration's (NASA) Public Use of Remote Sensing Data, this site contains many interesting applications of mathematics to the study of natural phenomena.

Autenfeld, Robert B. "W. Edwards Deming: The Story of a Truly Remarkable Person." Available on-line. URL: http://www.iqfnet.org/IQF/Ff4203.pdf. Downloaded June 2, 2003. This is a sympathetic and very detailed biography of Deming. It has a good deal of interesting information.

The British Museum. Available on-line. URL: http://www. thebritishmuseum.ac.uk/compass/. Updated June 2, 2003. The British Museum is one of the world's great museums. It has an extensive collection of images of ancient artifacts accompanied by informative captions. See, for example, the Royal Game of Ur, also called the Game of 20 Squares, from ancient Mesopotamia. A virtual version of this, one of humanity's oldest known games of chance, can be played at this site as well.

The Eisenhower National Clearinghouse for Mathematics and Science Education. Available on-line. URL: http://www.enc. org/. Updated on June 2, 2003. As its name suggests, this site is a "clearinghouse" for a comprehensive set of links to interesting sites in math and science.

Electronic Bookshelf. Available on-line. URL: http://hilbert. dartmouth.edu/~matc/eBookshelf/art/index.html. Updated on May 21, 2002. This site is maintained by Dartmouth College. It is both visually beautiful and informative, and it has links to many creative presentations on computer science, the history of mathematics, and mathematics. It also treats a number of other topics from a mathematical perspective.

Eric Weisstein's World of Mathematics. Available on-line. URL: http://mathworld.wolfram.com/. Updated on April 10, 2002. This site has brief overviews of a great many topics in mathematics. The level of presentation varies substantially from topic to topic.

Euler, Leonhard. "Reflections on a Singular Kind of Lottery Named the Genoise Lottery." Available on-line. URL: http:// cerebro.xu.edu/math/Sources/Euler/E812.pdf. This is one of Euler's own papers on the mathematics of lotteries. It begins easily enough, but it quickly demonstrates how difficult the calculations associated with this subject can become.

Faber, Vance, et al. This is MEGA Mathematics! Available on-line. URL: http://www.c3.lanl.gov/mega-math. Updated June 2, 2003. Maintained by the Los Alamos National Laboratories, one of the premier scientific establishments in the world, this site has a number of unusual offerings. It is well worth a visit.

Fife, Earl, and Larry Husch. Math Archives. "History of Mathematics." Available on-line. URL: http://archives. math.utk.edu/topics/history.html. Updated January 2002. Information on mathematics, mathematicians, and mathematical organizations.

Frerichs, Ralph R. John Snow Site. Available on-line. URL: http://www.ph.ucla.edu/epi/snow.html. Updated May 8, 2003. This site is associated with the University of California at Los Angeles Department of Epidemiology. It gives an excellent description of John Snow and the cholera outbreak at Broadstreet.

Gangolli, Ramesh. *Asian Contributions to Mathematics.* Available on-line. URL: http://www.pps.k12.or.us/depts-c/mc-me/be-as-ma.pdf. Updated on June 2, 2003. As its name implies, this well-written on-line book focuses on the history of mathematics in Asia and its effect on the world history of mathematics. It also includes information on the work of Asian Americans, a welcome contribution to the field.

Heinlow, Lance, and Karen Pagel. "Math History." Online Resource. Available on-line. URL: http://www.amatyc.org/Online Resource/index.html. Updated May 14, 2003. Created under the auspices of the American Mathematical Association of Two-Year Colleges, this site is a very extensive collection of links to mathematical and math-related topics.

Johnston, Ian. "The Beginnings of Modern Probability Theory." Available on-line. URL: http://www.mala.bc.ca/ ~johnstoi/darwin/section4.htm. Updated April 2003. Part of a much larger online installation entitled . . . *And Still We Evolve: A Handbook on the History of Modern Science*, this site, maintained by Malaspina University-College, is thorough and accessible.

The Math Forum @ Drexel. The Math Forum Student Center. Available on-line. URL: http://mathforum.org/students/. Updated June 2, 2003. Probably the best website for information about the kinds of mathematics that students encounter in their school-related studies. You will find interesting and challenging problems and solutions for students in grades K–12 as well as a fair amount of college-level information.

Melville, Duncan J. Mesopotamian Mathematics. Available on-line. URL: http://it.stlawu.edu/ca.dmelvill/mesomath/. Updated March 17, 2003. This creative site is devoted to many aspects of Mesopotamian mathematics. It also has a link to a "cuneiform calculator," which can be fun to use.

O'Connor, John L., and Edmund F. Robertson. The MacTutor History of Mathematics Archive. Available on-line. URL: http://www.gap.dcs.st-and.ac.uk/~history/index.html. Updated May 2003. This is a valuable resource for anyone interested in learning more about the history of mathematics. It contains an extraordinary collection of biographies of mathematicians in different cultures and times. In addition, it provides information about the historical development of certain key mathematical ideas.

Probabilistic safety assessment: an analytical tool for assessing nuclear safety. Available on-line. URL: http://www.nea.fr/html/ brief/brief-08.html. Updated April 2003. This site, maintained by the French Nuclear Energy Agency, gives an interesting and nontechnical overview of how probability theory can be used to enhance nuclear safety.

United States Census. Available on-line. URL: http://www. census.gov/. Updated June 2, 2003. The U.S. Census Bureau produces an astonishing number of detailed statistical reports. This is an excellent source of insight into how statistics is used in practical situations.

PERIODICALS, THROUGH THE MAIL AND ON-LINE

+Plus

URL: http://pass.maths.org.uk
A site with numerous interesting articles about all aspects of high school math. They send an email every few weeks to their subscribers to keep them informed about new articles at the site.

Chance

URL: http://www.stat.duke.edu/chance/
This on-line magazine describes itself as *The Scientific American* of probability and statistics. It is an excellent source of ideas and contains many entertaining articles.

Function

Business Manager
Department of Mathematics and Statistics
Monash University
Victoria 3800
Australia
function@maths.monash.edu.au
Published five times per year, this refereed journal is aimed at older high school students.

The Math Goodies Newsletter

http://www.mathgoodies.com/newsletter/
A popular, free e-newsletter that is sent out twice per month.

Parabola: A Mathematics Magazine for Secondary Students

Australian Mathematics Trust
University of Canberra
ACT 2601
Australia
Published twice a year by the Australian Mathematics Trust
in association with the University of New South Wales, *Parabola*
is a source of short high-quality articles on many aspects of
mathematics. Some back issues are also available free on-line.
See URL: http://www.maths.unsw.edu.au/Parabola/index.html.

Pi in the Sky

http://www.pims.math.ca/pi/
Part of the Pacific Institute for the Mathematical Sciences, this
high school mathematics magazine is available over the Internet.

Scientific American

415 Madison Avenue
New York, NY 10017
A serious and widely read monthly magazine, *Scientific American*
regularly carries high-quality articles on mathematics and mathematically intensive branches of science. This is the one "popular"
source of high-quality mathematical information that you will find
at a newsstand.

INDEX

Italic page numbers indicate illustrations.